# Homo Ex Data

## The Natural of the Artificial

**Edited by Burkhard Jacob, Vito Oražem, Peter Zec**

나건 | 이용혁 옮김

*reddot* edition

# Editor

**피터 젝 회장**
**Prof. Dr. Peter Zec**

Peter Zec 회장은 레드닷의 설립자이다. 국제 디자인 기구인 International Council of Societie
of Industrial Design(ICSID)의 회장(2005~2007)을 역임 후에 레드닷 디자인 어워드를 설립하였다.
2006년 10월 독일의 경제지 'Wirtschaftschoche'는 그를 창의적이고 파격적 사상가 20인' 으로
선정하였으며, 2016년에는 서울시의 명예 시민이 되었다.

**버크하르트 제이콥 원장**
**Burkhard Jacob**

2010년부터 피터 젝과 함께 레드닷 디자인을 이끌고 있으며, 현재까지 레드닷 디자인 협회를 운영과
함께 디자인 산업에 대한 변화들을 연구, 관찰하고 있다.  2010년 피터 젝 회장과 함께 'Design Value
- A Strategy for Business Success', 2018년 'The Form of Simplicity - Good Design for a better
Quality of Life', 그리고 2019년에는 'The Book of Possibilities'의 저술책이 있으며, 시대의 흐름과
산업의 흐름과 함께 디자인의 방향에 대한 연구를 이어나가고 있다.

**비토 오라젬 부사장**
**Vito Oražem**

2018년부터 레드닷의 부사장과 레드닷 박물관 이사직을 맡고 있으며, 미디어 과학, 미술 이론, 문학,
그리고 디자인 홍보의 경험으로 피터 젝 회장과 함께 레드닷을 다양한 분야와 접목 시키는데 이바지 하고
있다. 저서로는 2004년 피터 젝과 집필한 'Emerging Paradigm: Design and Change: Inventing New
Forms of Experience and Communication'이 있다.

## 옮긴이

**나건 교수**
**Ken Nah, Ph.D., CPE**

홍익대학교 IDAS(국제디자인전문대학원) 디자인경영 교수로 재직 중이며, Red Dot Award: Product Design 및 Design Concept 심사위원, 2023 광주디자인비엔날레의 총감독을 맡고 있다.

**이용혁 대표**
**Yong Lee**

현재 홍익대학교 IDAS(국제디자인전문대학원) 디자인학 박사과정 중이며, 미국에서 Fine Art를 전공하고 연세대학교에서 산업디자인학부, 동 대학원 디자인경영협동과정 석사를 졸업하였다.
현재 디자인 법인 오컴스와 브랜딩 스튜디오 찰리파커의 대표이사로 활동하고 있다.

Preface

# Homo Faber,
# to the Homo ex Data

**Ken Nah**

**Professor, IDAS, Hongik University | Red Dot Jury**

**Yong Lee**

**Ph.D Student, IDAS, Hongik University | CEO of Ockhams, Charlie Parker**

The history of mankind can be said to be a constant repetition of innovation. Innovation is only possible through trial and error. The first human, 'Homo Sapiens' was also able to develop by creating necessary tools and redesigning the environment through trial and error. Since the 18th century, we humans have created tangible and intangible tools through the $4_{th}$ industrial revolutions, and now even virtual reality tools.

In this book, Peter Zec, the founder of Red Dot, defines humanity as 'Homo Faber' from 'Homo Sapiens' in the era of the $4_{th}$ industrial revolution. As far as tools concerned, primitive stone hammers can also be tools, and complex cogs that driving steam engines in the 18th century can also be said to be tools. However, even a single simple word needs classification in certain times, and a new definition of humanity that adapts to the new environment is also needed. Dr. Peter Zec defined the human race that creates tools while living in the current Fourth Industrial Revolution as Homo Ex Data in this book. All services using metaverse, blockchain, AI, and smart phones that we currently encounter in every day are data-based, and the human using the data as tools is defined as Homo ex Data. If so, the question here may arise, "Did you advocate this ecological classification in the field of design, not in the field of science?" The reason is that design is a key element of disruptive innovation in the era of technology-led convergence. Therefore, the role of designers should be able to find new opportunities for innovation among the big data.

The purpose of translating this difficult book also came to understand the trend of the world and to be helpful to all designers who want design innovation. Once again, I would like to thank Peter Zec, the chairman of Red Dot, Vito Oražem, and Burkhard Jacob for allowing us to translate this meaningful books, and I hope we will be a little help for design in Korea.

October 31, 2022 at the Yemungwan Lab.

서문

# 도구의 인류에서
# 데이터를 사용하는 인류로.

**나건**
홍익대학교 국제디자인전문대학원(IDAS) 교수 | 레드닷 심사위원
**이용혁**
홍익대학교 국제디자인전문대학원(IDAS) 디자인학 박사과정 | 오컴스, 찰리파커 대표

인류의 역사는 끊임없는 혁신의 반복이라 말할 수 있다. 혁신은 실패와 시행착오를 통해야만 가능한 일이다. 최초의 인류 '호모 사피엔스'도 시행착오를 통하여 필요한 도구를 만들고 환경을 새롭게 디자인함으로써 발전할 수 있었다. 우리 인류는 18세기 이후 4번의 산업혁명을 통하여 유·무형의 도구를 만들어내고, 이제는 가상현실의 도구까지도 만들어 내고 있다.

이 책에서 '레드닷'의 설립자인 Peter Zec 회장은 4차 산업혁명과 동시에 인류는 '호모 사피엔스'에서 '호모 파베르(도구의 인류)'라고 정의하고 있다. '도구'라는 범주 안에는 원시 시대의 돌망치도 도구 일 수 있고, 18세기의 증기 기관을 움직이게 하는 복잡한 톱니바퀴 역시 도구라고 말할 수 있다. 하지만, 단 하나의 단순한 단어에도 시대적으로 분류가 필요하고 새로운 환경에 적응하는 인류의 새로운 정의도 필요하다. Peter Zec 회장은 지금의 4차 산업혁명을 살아가며 도구를 만들어내는 인류를 이 책에서 호모 엑스 데이터 (Homo Ex Data)라고 정의했다. 현재 우리가 매일 뉴스에서 접하고 있는 메타버스(Metaverse), 블록체인 (BlockChain), AI(Artificial Intelligence), 그 외 스마트폰을 이용한 모든 서비스가 데이터 기반의 산업이며, 이것을 도구로 사용하는 인류를 호모 엑스 데이터라고 정의한 것이다. 그렇다면 여기서 궁금한 것은 '과학 분야가 아닌 디자인 분야에서 이 같은 인류 생태학적인 분류를 주창하게 되었나?'라는 의문이 생길 수 있다. 그 이유는 디자인은 기술 주도의 융복합 시대에서 파괴적 혁신(Disruptive Innovation)의 핵심요소이기 때문이다. 따라서 디자이너의 역할은 빅데이터를 활용하여 새로운 혁신의 기회를 찾아낼 수 있어야 한다.

이 어려운 책을 번역하는 목적 역시 세상의 흐름을 이해하고, 혁신을 디자인하고자 하는 모든 디자이너에게 도움이 되고자 하는 마음에서 번역·출간하게 되었다. 원서 그대로의 늬앙스를 전달하기 위하여 다소 거친 번역체로 기술하였으니 영어 원문과 같이 이해하면 도움이 될 것이라 생각한다.
다시 한번 의미 있는 책의 번역을 허락해 준 레드닷의 Peter Zec 회장과 관계자들에게 고마움을 표하며, 국내 디자인 발전에 자그마한 도움이 되길 바란다.

2022년 10월 31일 예문관 연구실에서.

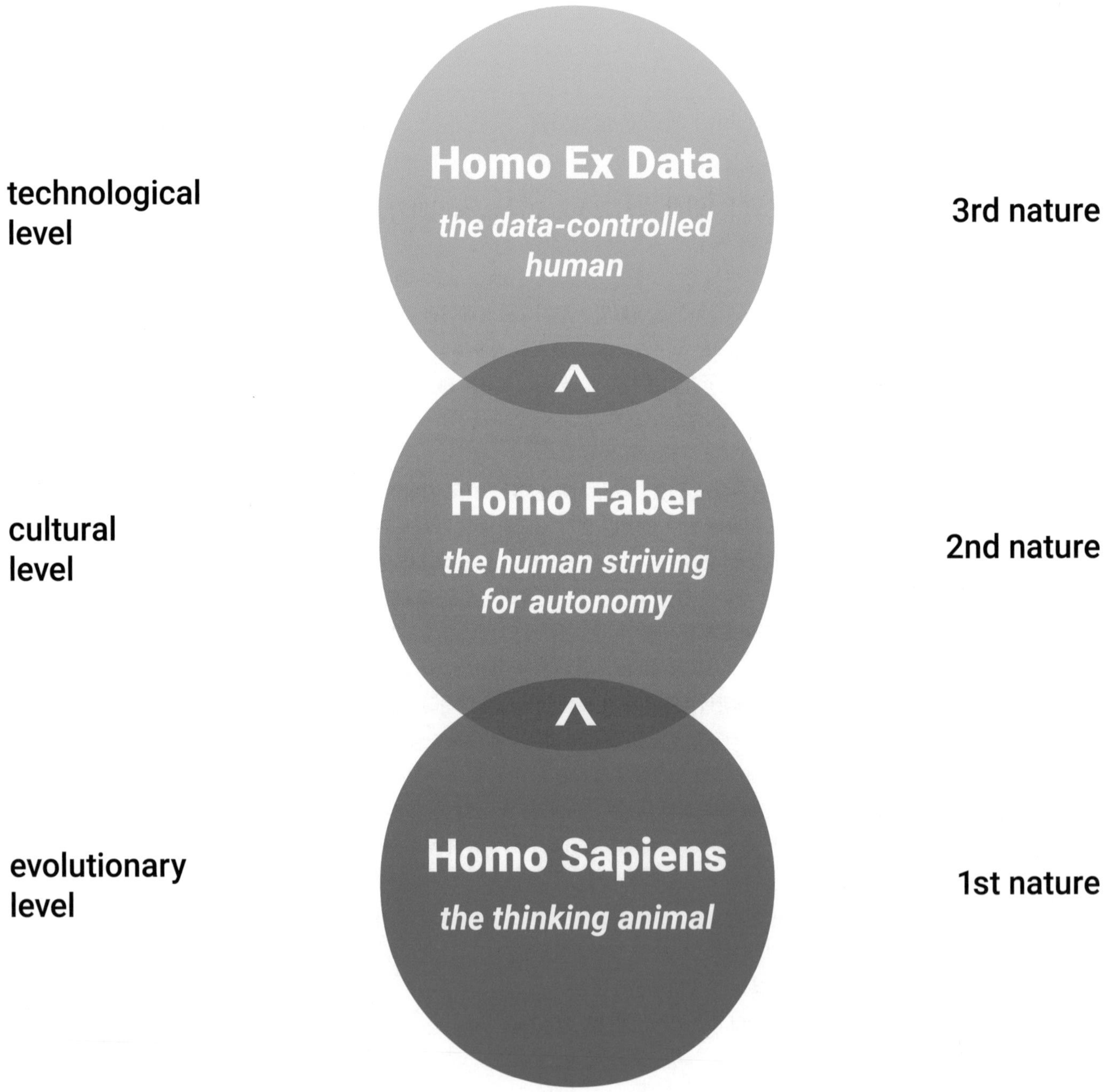
technological level
Homo Ex Data
the data-controlled human
3rd nature
^
cultural level
Homo Faber
the human striving for autonomy
2nd nature
^
evolutionary level
Homo Sapiens
the thinking animal
1st nature

Essay

# Homo Ex Data –
# The Natural of the Artificial

**Design in the Rising Age of "Big Data"**
**by Professor Dr Peter Zec, founder and CEO of the Red Dot Award**

## Homo Sapiens

With the modern-day, mass-production-based second industrial revolution, design is taking on the function of a creative quality at the interface between what is technically feasible and what is humanly manageable. Design therefore makes a significant contribution to how humans identify with their new surroundings and the objects available to them in these surroundings. However, whether or not the things around us are visually appealing is, at first, entirely irrelevant. In the relationship between humans and the surroundings that they create artificially, it is not the beauty of the artificial world that is most important, but the skill and proficiency in using things in interaction with the artificially created reality.

Design gives us not only the feeling, but – when successful – also the certainty that we are in control of things. What is reflected in this relationship is, not least, also a specific form of the will to have power over life and our surroundings. This is ultimately what distinguishes modern humans from their prehistoric ancestors, whose lives were still solely determined by the will to survive in nature. Today, we commonly refer in this connection to "first nature", which early humans had to face virtually without any power, literally "powerlessly". Nevertheless, prehistoric humans from early cultures did not simply resign themselves to a helpless fate. They instead developed creative skills and dynamic imaginations in their daily struggle for survival when confronted with the challenges and dangers of their natural surroundings.

It is this early phase of human history that saw the emergence of *Homo sapiens*, the first life form to develop a certain insight into and understanding of their surroundings and to create tools that made it easier for them to survive in nature. These things – prehistoric tools and weapons – initially stemmed from a feeling of powerlessness in the face of the natural surroundings. *Homo sapiens* also developed the first modes of behaviour and rituals, which served to give structure and order to a type of family or social community. They also helped counterbalance inexplicable phenomena and the ensuing feeling of being powerless in the face of nature. As humankind has continued to develop, the various rituals have undergone continuous change, with some disappearing entirely, and others being replaced by new forms. So, we can safely say that making tools as well as devising and practising rituals has remained a key feature of the development of human history right through to the present day.

In the prehistoric objects that *Homo sapiens* created, design did not play even a minor or insignificant role. It was simply not even a consideration. Things initially came about coincidentally, and later some forethought was given, but always from a position of powerlessness. Design, however, requires deliberate awareness as well as endeavouring to gain power over how things are used. At the same time, design perpetually aims to optimise what already exists. It is a constant battle to replace good things with better things. When we look at it this way, design was only able to come about when humans had gained power over nature and their surroundings and created a kind of "second nature" by means of technology.

»Design can only come about when humans have gained power over nature and their surroundings and create a kind of "second nature" through technology.«

Unlike in the world of art, the main premise of design is not aesthetics, but rather the appropriateness, expediency, and utility of an object. This is one of the reasons why the German design historian Gert Selle argues: "In pre-industrial times, there was no postulating design that would have insisted on a use that would not have been 'obvious' as an internalised socio- and psychogenetic necessity."[1] The development of design first emerged as an independent discipline with advancing industrialisation. The development of industrial manufacturing conditions was accompanied by a fundamental departure from the handcrafted production methods that had been cultivated up to that point. At the same time, people's everyday lives changed at a seemingly dramatic pace. The rhythm of day and night dictated by nature and the pattern of working and sleeping that resulted from it played out ad absurdum in the newly created factory halls. Artisanal skills were replaced by monotonous and stereotypical, machine-based work on the production line. This development resulted in masses of products manufactured in series, which were rapidly distributed throughout the newly emerging industrial society.

## Homo Faber

The mass production of serial products thus became the point of departure for the emergence of an entirely new kind of social reality. According to Gert Selle, "Product forms reflect production methods and ways of living."[2] Accordingly, as a result and a necessity of this development, demand was created for ever-newer and, above all, better product forms, which, reduced to a functional form determined by a machine, became estranged from manual craftsmanship. After all, a machine can only produce the products that it is able to make. If the production sequence in the machine changes, the product that emerges is also altered. The new challenge of this era was then to create machine-based products that improved people's lives. In order to do this, there was a constant need for new product ideas expressed in tangible design forms. At the same time, it goes without saying that the machines also had to be modified and improved accordingly each time so as to realise these new product forms.

**»Homo faber designs and produces tools and machines in order to constantly manufacture new products that make people's lives easier and, ideally, even enrich their lives.«**

This transformation of life in the on-going development of modern society shaped by industrial production gave rise to a whole new type of human, *Homo faber*. *Homo faber* designs and produces tools and machines in order to constantly manufacture new products that make people's lives easier and, ideally, even enrich their lives. In this way, *Homo faber* endeavours to actively shape, cultivate, and manage his surroundings. Nonetheless, on their own, an on-going striving for technical innovation and continuous optimisation of production processes did not suffice to make products capable of improving the quality of everyday life. What was now necessary was to put what was technically feasible in a usable form. Technology therefore needed to be combined with a socially applicable and usable form. This became the new task of design.

Of course, every artificially created object has a specific form that expresses its unique nature and special aspects. This also applied equally in the case of the first prehistoric tools made by humans. The understanding of form and related processes in pre-industrial societies, however, differed decisively from that of the design that emerged as part of industrialisation. Originally, the maker of an object was more or less free to determine the form of the object through his or her personal handicraft skills. The related understanding of form resulted from the respective social reality. This was shaped in turn by the prevailing tastes of the time as well as by ritualised conventions, customs, and habits. In the pre-industrial era, the design of objects was therefore primarily defined by societal rituals and the individual skills of a master.

Industrial mass production was inevitably accompanied by a fundamental shift in social reality and prevailing ideas and values. With the help of technology, it seemed possible for the first time in human history to liberate humans from the supremacy of nature and allow them to take charge of their own destiny. This consequently gave rise to efforts to replace nature with technology in almost all areas of life. The task of design was to give this endeavour a new form. In the course of this process, design developed into its own aesthetic discipline, which continues to play a decisive role in the emergence and shaping of social reality to this day. The design of things is a manifestation and expression of our modern-day culture. At the same time, design is therefore also closely connected to society's technical and economic developments.

## »The design of things is a manifestation and expression of our modern-day culture.«

Design is created by people for people. People are consequently the yardstick for how things are designed. Up to now, this yardstick has been based on a striving for autonomy and a better quality of life, with the will to have power over nature and to live a self-determined life playing a central role in this regard. Technology and design serve as tools for asserting this striving for power.

For a time, the newly emerging reality of life was experienced as an act that freed human beings from the threats, burdens, and dangers of nature. This development gave rise to euphoria and boundless self-confidence in the autonomous skills of humanity. Technology was then increasingly used to oust nature from human reality, which means that, today, the modern environment of human beings is shaped predominantly by artificially created things. Nevertheless, since people continue to regard themselves as the yardstick for the things that surround them, nature still always plays a role when it comes to designing a new reality. It is not possible to ignore or deactivate human nature in this equation.

## Function and Use

The advancing development of technology makes it possible over time to create more and more perfect and more and more complex artefacts. The task of design today is therefore to design things in such a way that they remain manageable for humans going forward. The focus here is on use. The designer Otl Aicher goes so far as to describe use as "a new criterion of truth"[3] for the design of everyday things. The aesthetics of design differ quite considerably from the aesthetics of art in this regard. Unlike art, which can be entirely arbitrary and completely independent of its surroundings, there is no absolute design aesthetic per se. Design is, consequently, always geared to the actual interaction with things.

Design does not have to be conceived from an abstract context and justified. In fact, just the opposite is the case. Design is manifested solely through the act of making. How things should look "should no longer be determined by form, the aesthetic principle, but by use. Form does not result from a code, however clear, not from art, but from application."[4] The form of things thus stems from the specific way of life in interacting with things. "Use is the renunciation of everything that wants to explain. The thing itself expresses itself in its use."[5] The quality of the design flows automatically from use. Poorly designed things are poorly designed because they are not easy to use. Conversely, well-designed things are easy to use. But how can the quality of use be defined clearly and unambiguously? What one person considers good and easy to use is not necessarily the same for others. This means that we have to adopt a much more nuanced approach when assessing the form of use. It must transcend the general form of interaction of any one person.

In his deliberations on the form of things, the U.S.-American computer scientist Herbert Simon goes beyond merely assessing use. He defines the functional or purposeful aspect of things as an essential feature for distinguishing the artificial from the natural. People produce artificial things in order to fulfil a certain purpose with them. Simon distinguishes between what he calls an "inner" and an "outer" environment, which converge in the artefact as an "interface". "An artefact can be thought of as a meeting point – an 'interface' in today's terms – between an 'inner' environment, the substance and organization of the artefact itself, and an 'outer' environment, the surroundings in which it operates. If the inner environment is appropriate to the outer environment, or vice versa, the artefact will serve its intended purpose."[6] We could also describe the inner environment as the function of the artefact. This is generally geared towards a purpose or an objective. In a way, the outer environment of the artefact corresponds to its use. However, this perspective also takes into account the surroundings in which the artefact is supposed to perform. This broader view restricts and relativises the importance of use. For example, the function and design of an aeroplane are geared towards making the artefact fly. But this does not achieve the entire objective, since the aeroplane also has to accomplish certain objectives on the ground. It must be able to manoeuvre at airports in such a way as to prevent any problems with the surroundings from arising. On the other hand, the design of the airport must also allow for aeroplanes to be manoeuvred easily on the ground. For use on roads, however, aeroplanes are just as unsuitable as roads are for aeroplane transportation.

The cockpit of an aeroplane must be designed to allow a pilot to navigate it easily. It is not sufficient for all displays and operating instruments to function with technical perfection. The pilot must also feel safe when using the instruments. Human error is still the cause of between 70 and 80% of all aeroplane disasters. In the highly complex surroundings of an aeroplane, a human being cannot retain 100% control. There are, however, no statistics on how often pilots have prevented crashes after technical failure thanks to their individual expertise. Trusting in the pilot's expertise is very reassuring for most passengers. Even if it were possible to fly planes without pilots, many travellers would surely feel uncomfortable about it or would even choose not to fly anymore. We are simply used to putting our trust in the special skills of human beings. Nevertheless, we can observe how humans are increasingly disappearing from technically complex systems as active participants. This will and must have a hugely significant influence on the design of such systems.

**»We can observe how humans are increasingly disappearing from technically complex systems as active participants.«**
The technical environment has long since become a kind of second nature to us. The driving force behind this is the development of technical innovations that are better and better and more and more efficient and complex. Design always tries to keep pace with this development, something that is becoming more and more challenging in view of the dynamic pace of innovation. The aim of design continues to be making the technical environment manageable and

putting it at the service of human beings. In the same way that humans once used technology to emancipate themselves from nature and control it as far as possible, design now serves to control technology. To allow humans to remain in charge of the environment that they themselves have created, numerous products and processes will have to be radically simplified. Simplicity in all areas of life has now become one of design's most important goals. Simplicity in design can be achieved in a variety of different ways: by making objects, user interfaces, systems, and components smaller or larger. Reducing the operating options is also an effective means of achieving simplicity. The use of things must remain manageable. Otherwise, people will quickly start to feel uneasy.

## Emotions

Although humans have subjugated nature by means of technology, they themselves are and remain natural beings and thus a fixed part of nature. Most technical inventions are also related to nature in varying degrees, because they too must respect the laws of nature in order to achieve their objectives. It is people who, in turn, always formulate these objectives. They generally serve to satisfy the natural requirements of human beings. "As our aims change, so too do our artefacts, and vice versa."[7] Each of these aims is based on a motivation, which, in turn, is based on emotions. The biologist Humberto Maturana describes emotions as a fundamental condition of human nature. "Emotions are dynamic body dispositions that specify the domains of actions in which animals in general, and we human beings in particular, operate at any moment."[8]

This means that all human action – and thus also formulating goals – has its basis

**»Emotions are dynamic body dispositions that specify the domains of actions in which animals in general, and we human beings in particular, operate at any moment.«**

in emotions. Unlike humans, machines and artificial systems as well as all kinds of artefacts do not operate in a domain of action that is specified by an emotion. Even machines controlled by artificial intelligence are not yet able to motivate themselves to act, since they are not guided by emotions. From today's perspective, it is scarcely imaginable that machines will ever be able to reach this natural stage.

So what do these findings mean for the design of present-day and future artefacts such as robots or intelligent, driverless transport systems, aircraft, and the like? Regardless of how utopian these innovations may appear to us, it is still humans who formulate and implement the related objectives. Designers develop strategies that serve to equip artificial things with more natural features. In this respect, nature serves not only as a source of inspiration for innovative forms, but also as a stylistic design tool. But the greatest challenge is still the emotional aspect.

While artefacts are entirely free of emotions, human actions are determined to a large extent by emotions. This is equally true for interaction with artefacts. Regardless of whether we are standing in front of the refrigerator, sitting in front of a laptop, or controlling a robot, our actions are accompanied by emotions at any given time. We can feel joy, sympathy, indifference, or even unease or fear without the artefacts that surround us having the least idea of what we are feeling. This is also one of the main reasons why designers try to design objects in such a way that they evoke emotions in us that are as positive as possible. This applies in particular to objects that we use very individually and every day. For example, personal robots are often designed to look like a small child in order to evoke an emotional connection on the part of the user. Furthermore, giving objects childlike qualities makes them appear harmless, helping users not to be afraid of what appears to be a perfectly autonomous machine. In this way, artefacts without any emotions are practically perceived as living beings.

Aeroplanes are not artificial birds. They move in an entirely different way, but are nonetheless able to fly. In the same way, robots move differently from humans, but are able to carry out numerous tasks for us. The original plan was even for robots to replace humans. As early as 1920, the Czech author Karel Čapek's drama "R.U.R. – Rossum's Universal Robots" saw a robot heralding the end of humankind: "Robots of the world! Many humans have fallen. We have taken the factory and we are masters of the world. The era of man has come to its end. A new epoch has arisen! Domination by robots!"[9] At the time, the second industrial revolution was in full swing. Production lines and electrification had completely changed factory work and made mass production possible. Numerous critics attacked the new working conditions on the production line, calling them inhuman. The production line workers became part of an automated process, with their tasks reduced to a few monotonous hand movements. This situation was ultimately also highlighted and criticised in numerous books, plays, and films. Čapek's drama can, consequently, also be regarded as a reaction to these working conditions. The monotonous and simple work steps on the production line were instrumental to the idea that artificial workers (robots) would replace human beings.

## The Natural and the Artificial

What was once a utopian fantasy has long since become an everyday reality. As mentioned at the outset, the spread of mass production was accompanied by the emergence of design as a new discipline in the creation of industrially manufactured products. At the time, design pioneers very quickly recognised that design should not be restricted to mere superficial beautification of the machine-produced goods like a form of product cosmetics. From the very beginning, the objective of designers was geared toward having a direct influence on the entire product development process. Designers and engineers were to work together to come up with the best solution for a new product. Today, this idea is accepted without question, but, in the past, many entrepreneurs did wonder whether it made any sense, considering it divorced from reality. But what role will design play in future, when products are only conceived by humans but developed and designed by machines using biomimicry and artificial intelligence?

In his book "Out of Control", U.S. journalist Kevin Kelly posits that machines are becoming biological and the biological is becoming engineered. He describes this development as "the marriage of the born and the made".[10] The natural and

>>The natural and the artificial will combine to form a new "bio-artificial" reality.<<

the artificial will combine to form a new "bio-artificial" reality. This development will result in a new living situation in which humans work and live in a world with living artefacts. At the same time, this will also involve humans losing control of the world. We can already see the point of departure for this specific utopia in the past. "For the world of our own making has become so complicated that we must turn to the world of the born to understand how to manage it. That is, the more mechanical we make our fabricated environment, the more biological it will eventually have to be if it is to work at all."[11] As a result, humanity is headed toward a new technological future, which Kelly refers to as a "neo-biological civilization". This path will run in two opposite directions, which Kelly describes as follows: "Human-made things are behaving more lifelike, and life is becoming more engineered."[12]

What are the consequences if things behave in an ever more lifelike way? Firstly, it means that they are in a position to organise themselves, learn, and change themselves in order to bring about their own shared reality. To do so, they have to be able to carry out effective actions. This skill is based, in turn, on communication and recognition, in other words, on the development of a specific form of intelligence. Nevertheless, this does not mean that artefacts will come anywhere close to biological beings, since their behaviour – regardless of how lifelike it may seem to us – is based not on emotions but on "coordinators". "The basic idea is that a coordinator replaces standard e-mail with software that represents the discussion and interaction relationships constantly taking place in an organisation and makes them accessible. Coordinators are just one example of the reorientation of artificial intelligence: instead of asking computers to do something that is probably impossible, e.g. the never-ending task of linguistic analysis, we use it instead as a structuring medium for explicitly recording and presenting our on-going linguistic creation of the world."[13]

## Homo Ex Data

But what does it mean for humans if their lives are becoming more engineered? Kevin Kelly describes this as technology being "out of the control" of humans. In addition, humans' entire surroundings are changing. In the same way that things can already "communicate" and "interact" with other things today, the same applies with respect to the relationship between us humans and the artefacts that we equip with artificial intelligence. It is no longer about designing individual products or objects, but about creating a new environment and reality for life.

At the same time, we humans form and control this process less and less. In a best-case scenario, we can still speak in some situations of a co-production between humans and technology or between the natural and the artificial. In this way, our surroundings become a network in which an indecipherable number of sensors register an even more indecipherable amount of data, which in turn generates a new environment. In this new world, we are subjected to a continuous optimisation process at every turn, in the truest sense of the word. As consumers, we learn from products in the same way that products also learn from us as users. The world and our surroundings are therefore becoming a construct of on-going, sensor-controlled data transfer.

These events inevitably also have a substantial impact on human development. Following *Homo sapiens* and *Homo faber*, a new type of human, which we call *Homo ex data*, is emerging: human beings who no longer base their existence solely on a shared biological tradition or on controlling their environment by means of technology, but whose living circumstances are instead determined by the generation and transfer of data. *Homo ex data* is no longer actively involved in shaping his surroundings. Instead, his actions become more and more dependent on large-scale data collection and evaluation.

It is increasingly uncommon for products to be developed based on observations and experience regarding function and use. They instead result from

**»*Homo ex data* is no longer actively involved in shaping his surroundings. Instead, his actions become more and more dependent on large-scale data collection and evaluation.«**
an optimisation process based on comprehensive analysis of specific data. The design of products is also not able to escape this dominance of data transfer. What ensues is a data-relevant logic of function and user guidance. In extreme cases, the only escape is as an aesthetic accessory. Things of the era of *Homo ex data* conform to the system. They often only differ very slightly in terms of their appearance. The actual design is shifted to the interior of things. Function, purpose, and use are determined and controlled by sensors. Errors are immediately "communicated" to the network. The optimisation process is activated right away and processed for an update. Design diversity is now an exception rather than the rule.

The production of things for everyday life is also shifting to the download of data to control a digital production process. This, however, also presents an opportunity for *Homo ex data* to engage in a type of co-creation by having his individual data flow directly into the product design. Products are integrated with other products and with information from the environment and the user, which are collected and analysed. The existence of *Homo ex data* is monitored, observed, and evaluated by a ubiquitous system of sensors, so that life is determined by universal data transparency and on-going data analysis. This is the phenomenon generally described by the term "Big Data". It refers to a gigantic mass of data volume that is collected, analysed, and evaluated for any given situation at an incredible speed and with an immeasurably wide range of data sources. Everyone and everything can be used as a data source provided that the environment and the living creature – the natural and the artificial – are equipped with omnipresent sensors.
The U.S. author Dave Eggers' best-selling novel "The Circle"[14], which has also been made into a film, depicts a terrifying scenario of a future in which society is dominated by total transparency, surveillance, and data control.

With *Homo ex data*, human evolution is entering a whole new phase. Unlike *Homo faber*, this new type of human is no longer interested in having power over technology in order to control his environment. Instead, such humans themselves become part of a world that strives for absolute transparency, organised by universal data transfer. This world regenerates itself again and again by coupling artificial with natural systems. Fitted with and surrounded by sensors, *Homo ex data* becomes part of a systematically constructed, self-referencing reality. The life of *Homo ex data* is very different from that of *Homo faber*, as subjectivity and individuality are experienced and assessed in a completely different way. Possession and ownership for self-realisation become less important in favour of a universally available offering of services.

The design of these different services meanwhile determines much more about how quality of life is experienced than the design of individual products. The context in which a product operates obtains greater relevance and

**≫With *Homo ex data*, human evolution is entering a whole new phase. Unlike *Homo faber*, this new type of human is no longer interested in having power over technology in order to control his environment. Instead, such humans themselves become part of a world that strives for absolute transparency, organised by universal data transfer. This world regenerates itself again and again by coupling artificial with natural systems.≪**

importance than the product itself. The significance of design shifts from the design of an individual product to the generation of complex systems in which *Homo ex data* becomes a part of the whole. At the same time, production and ownership conditions constantly change in this society. Mechanical forms of production are increasingly replaced by digital or digitally controlled systems. This allows for greater flexibility without resulting in a stronger emphasis on diversity. The task of design in this new situation is to design the "communication" and "interaction" of systems with systems as well as between humans and systems. The design of individual products loses significance in favour of the design of comprehensive services.

Like *Homo ex data*, the new *products ex data* must also function in a larger context. And that is the new challenge for design. In this world characterised by data transparency, products are no longer made solely by people for people. Unlike what Dave Eggers attempts to portray in his novel, however, the transition of humans and society described here is currently not linked to any general risk for humanity as long as the interaction of artefacts does not take place "in a domain of action specified by emotion".[15] Although the natural and the artificial are becoming more and more closely interlinked, so far, it is still only humans who are able to create emotionally influenced motivation.

[1] Gert Selle, "Designgeschichte in Deutschland. Produktkultur als Entwurf und Erfahrung" (Cologne, 1987), p. 20.
[2] Gert Selle: loc. cit., p. 7.
[3] Otl Aicher, "Analogous and Digital" (Hoboken, NJ, 2015), p. 88.
[4] Otl Aicher: loc. cit., p. 91.
[5] Otl Aicher: loc. cit., p. 92.
[6] Herbert A. Simon, "The Sciences of the Artificial" (Cambridge, 1969), p. 6.
[7] Herbert A. Simon: loc. cit., p. 3.
[8] Humberto R. Maturana, "Biologie der Realität" (Frankfurt, 1998), p. 325.
[9] Quoted from Red Dot 21. www.red-dot-21.com.
[10] Kevin Kelly, "Out of Control: The Rise of Neo-Biological Civilization" (New York, 1994), p. 2.
[11] Kevin Kelly: loc. cit., p. 2.
[12] Kevin Kelly: loc. cit., p. 3.
[13] Francisco J. Varela, "Kognitionswissenschaft – Kognitionstechnik" (Frankfurt, 1990), p. 114.
[14] Dave Eggers, "The Circle" (New York, 2013).
[15] Humberto R. Maturana: loc. cit.

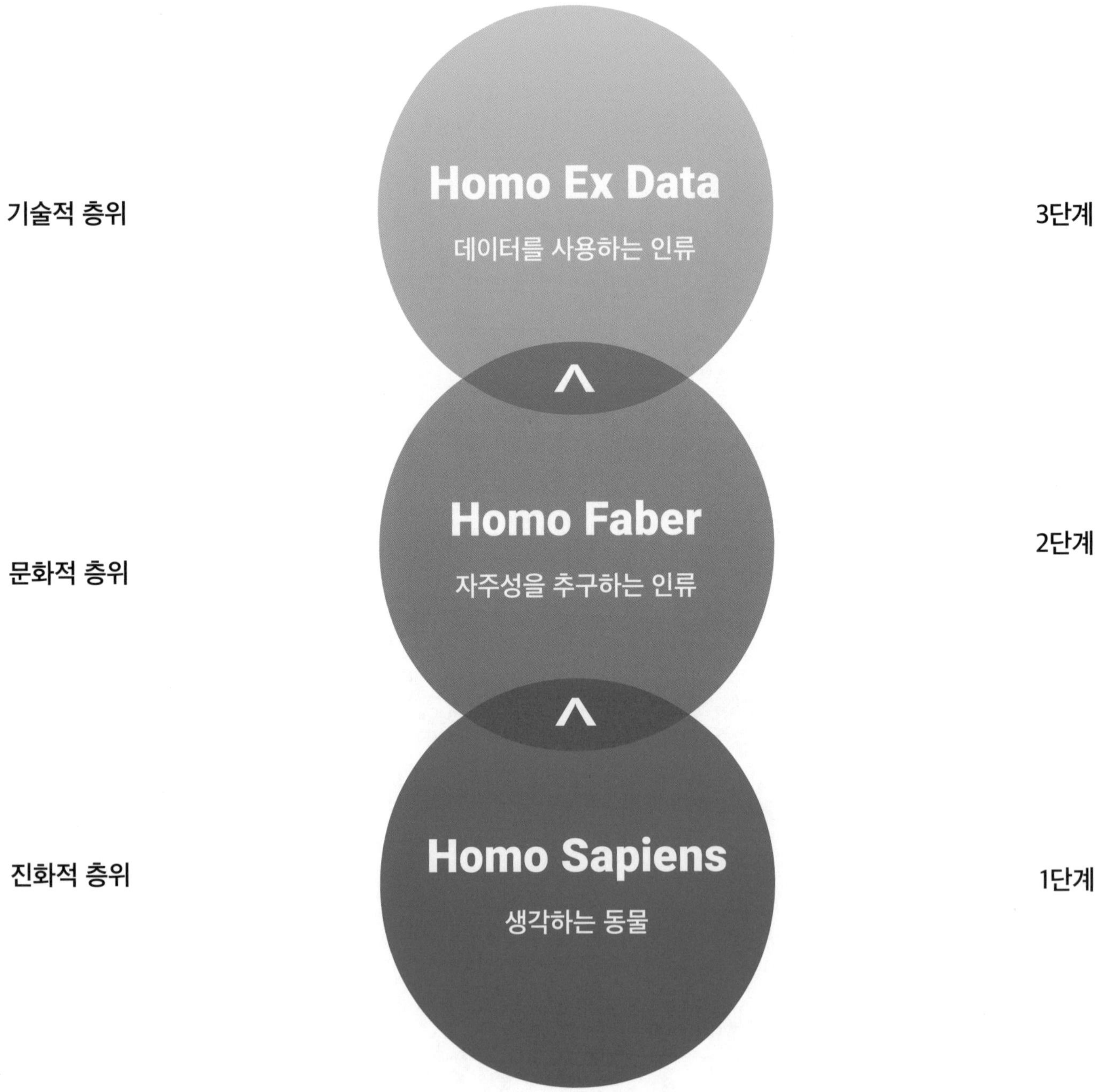

기술적 층위

**Homo Ex Data**

데이터를 사용하는 인류

3단계

∧

**Homo Faber**

자주성을 추구하는 인류

문화적 층위

2단계

∧

**Homo Sapiens**

생각하는 동물

진화적 층위

1단계

에세이

# 호모 엑스 데이터,
# 인공적 자연

Design in the Rising Age of "Big Data"
피터 젝, 교수 & 레드닷 어워드 회장

## 호모 사피엔스

현대의 대량 생산 기반인 2차 산업혁명은 디자인과 함께 기술의 가능 범위와 인간 사이의 경계에서 창의적 기능을 수행하고 있다. 디자인은 인간의 새로운 환경에서 사용할 수 있는 대상을 식별하는 방법에 상당한 기여를 하고 있지만, 보여지는 모든 것들이 시각적으로 매력적인지 아닌지는 알 수 없다. 인간이 만들어낸 주변 환경에서 가장 중요한 것은 인공적 환경의 아름다움이 아니라, 인간이 창조한 현실과 상호작용하여 사물을 사용하는 기술과 숙련도라고 말할 수 있다.

디자인이 성공적인 경우, 사용자는 사물을 제어하고 있다는 느낌과 확신을 받게된다. 이 관계에 반영된 것은 무엇보다도 생명과 주변 환경을 지배하고자 하는 특정한 형태의 의지이기도 하다. 이것은 현대인과 선사시대 인류를 구별 짓는 궁극적인 차이점이다. 오늘날 우리는 이와 관련하여 초기 인류가 사실상 아무런 힘도 없이, 문자 그대로 무력하게 직면해야 했던 '제1의 자연'을 일반적으로 언급한다. 그럼에도 불구하고 초기 문화의 선사 시대 인류는 무력한 운명에도 스스로를 포기하지 않았다. 자연환경의 도전과 위험에 직면했을 때 생존을 위한 매일의 투쟁을 통해 창의적인 기술과 역동적인 상상력을 개발했다.

호모 사피엔스의 출현을 본 것은 인류 역사의 초기 단계이며, 환경에 대한 특정한 통찰력과 이해를 쌓아 자연에서 더 쉽게 생존할 수 있도록 도구를 개발한 최초의 생명체이다. 선사 시대의 도구와 무기는 처음에는 자연환경에 대한 무력감에서 비롯되었다고 볼 수 있지만, 다른 한편으로는 가족이나 사회 공동체 유형에 구조와 질서를 발전시키고 번성하게 되었다. 그리고 무엇보다 이들이 설명할 수 없었던 자연 앞에서의 무력하다는 느낌을 상쇄시키는 데 도움이 되었다. 인류 문화의 발전에 따라 다양한 의식은 계속해서 변화해 왔으며 일부는 완전히 사라지고 일부는 새로운 형태로 대체되었는데, 도구를 만들고 의식을 수행한다는 것은 오늘날에 이르기까지 인류 역사 발전의 핵심 요소로 남아 있다고 자신 있게 말할 수 있다.

호모 사피엔스가 선사시대 창조한 물건들은 제대로 기능하지 않거나 하찮은 것들이었을지라도 모든 것들은 처음 우연하게 생겨났고, 효율적으로 기능하는 것을 알게 되었을 것이다. 그리고 이렇게 우연으로 알게된 도구는 더 나은 방법으로 개선을 통하여 최적화 되었을 것이다. 디자인은 좋은 것을 더 좋은 것으로 대체하는 것의 끊임없는 싸움이다. 이러한 관점으로 볼 때, 인류는 인간과 주변 환경을 이용하여 기술을 얻는 '제 2의 자연'을 창조할 수 있었던 것이다.

≫디자인이란, 인간이 주변 환경을 이용하여 기술을 얻는 '제 2의 자연'을 창조하는 것이다≪

디자인은 예술과 달리 기본 전제가 심미성 외에도 대상의 적절성, 편리성, 효용성이다. 독일의 디자인 역사가 Gert Selle는 이에 대해서 주장하였는데, "산업화 이전 시대에는 '명백한' 기능만을 보여주는 예상 가능한 디자인은 없었다."[1]고 정의했다. 산업혁명 제조 여건의 발달은 그 당시까지 계속되어온 수공 생산 방식에서 근본적으로 탈피하는 것을 의미했고 동시에 사람들의 일상은 극적인 속도로 변화하게 되었다. 자연이 지배하고 있는 낮과 밤의 리듬은 인간들이 지은 공장에서 일과 수면의 패턴이 새롭게 정립되는데, 수작업을 기반으로 하고 있던 장인의 기술은 생산 라인에서 단조롭고 틀에 박힌 기계기반 작업으로 대체되었다. 이러한 발전으로 인해 대량의 제품이 직렬로 제조되어 새롭게 부상하는 산업 사회 전반에 빠르게 보급되었다.


## 호모 파베르 (도구의 인류)

따라서 직렬생산라인의 대량 생산은 완전히 새로운 종류의 사회적 현실의 출현을 위한 출발점이 되었다. 게르트 젤레에 따르면 "제품의 형태는 제작 방식과 생활 양식을 반영한다." 이에 따라 이러한 발전의 필요성과 결과로 기계에 의해 결정되는 기능적 형태로 축소되어 수작업보다 새롭고, 무엇보다 우수한 제품 형태에 대한 수요가 창출되었다.

기계는 자신이 만들 수 있는 제품만을 생산할 수 있다. 기계의 생산 순서가 변경되면 나오는 제품도 변경된다. 이 시대의 새로운 도전은 사람들의 삶을 개선하는 기계 기반 제품을 만드는 것이었다. 이를 위해서는 실질적인 디자인 형태로 표현된 새로운 제품 아이디어가 끊임없이 필요했다. 동시에 이러한 신제품의 형태를 구현하기 위해 기계도 매번 그에 따라 수정되며 발전하였다.


≫호모 파베르는 사람들이 더 편하게 살고 이상적으로는 삶을 풍요롭게 하는 새로운 생산물을 지속적으로 만들기 위해 도구와 기계를 디자인하고 생산한다.≪

산업 생산에 의해 형성되는 현대 사회의 지속적인 발전에서 이러한 삶의 변화는 완전히 새로운 유형의 인간인 호모 파베르를 탄생시켰다. 호모 파베르는 사람들의 삶을 더 쉽게 만들고 이상적으로는 삶을 풍요롭게 하는 신제품을 지속적으로 생산하기 위해 도구와 기계를 설계하고 생산한다. 이처럼 호모 파베르는 자신의 주변 환경을 적극적으로 형성하고, 가꾸고, 통제하기 위해 노력한다. 그럼에도 불구하고 기술 혁신을 위한 지속적인 노력과 생산 공정의 최적화만으로는 일상의 질을 향상시킬 수 있는 제품을 만들기에 충분하지 않았다. 필요한 것은 기술 발전을 통해 사용 가능한 형태로 만들고, 발전된 기술은 사회에 귀속되었다. 그리고 이것이 디자인의 새로운 역할이 되었다.

물론 도구의 창작은 특별한 형태와 기능으로 생태계를 변화시키는 매개체가 되었는데, 이것은 인류가 만든 최초의 선사시대 도구에서부터 동일하게 적용되었다고 볼 수 있다. 그러나 산업화부터는 디자인의 새로운 기능이 부각되었고, 도구를 통해 제품을 제작하는 창작자들이 자유로워졌고 사회적으로 인정받게 되었다.

산업적 대량생산은 불가피하게 사회의 변화와 지배 사상의 근본적인 변화를 몰고왔다. 기술의 도움으로 인류는 역사상 처음으로 자연의 위대함에서 해방시키고 스스로의 운명을 스스로 개척할 수 있게 되었다. 결과적으로 거의 모든 삶의 영역에서 자연을 기술로 대체할 수 있었다. 여기서 디자인은 이러한 환경변화를 이루게 하는 주된 역할을 했다고 볼 수 있다. 이 과정에서 디자인은 전문 분야로 발전했으며, 이는 오늘날까지 인류 환경에 중요한 역할을 계속하고 있다. 사물 디자인은 현대 문화의 발현이자 표현이고, 디자인은 사회 기술 및 경제 발전에도 밀접하게 영향을 주게 되었다.

≫사물을 디자인한다는 것은 현대 문화의 발현이자 표현이다≪

디자인은 사람을 위해 사람이 만든다. 결과적으로 현재의 인류는 사물이 어떻게 설계되었는지에 대한 척도라고 볼 수 있다. 이 척도는 인류가 자연을 다스리고자 하는 의지와 자기주도적 삶을 살고자 하는 의지의 발현이라고 볼 수 있으며, 기술과 디자인은 인류의 권력 추구를 위한 도구로서의 역할을 하고 있다고 말할 수 있다.

오랜 시간 동안 인류는 자연의 위협으로부터 벗어나려는 시도를 통해 자주적인 기술 발전과 무한한 자신감을 얻게 되었다. 그 후 기술, 혹은 과학은 인류가 자연을 극복하는 요소가 되었으며 자신들의 환경을 주도적으로 창조한 인공적 요소로 형성되었다. 아직까지도 자연은 인간들에게 새로운 창작을 하게 되는 중요한 역할을 하고 있으며, 자연을 극복하는 것과 창작의 영감을 받는 대치적인 관계는 필수불가적이라 말할 수 있다.

## 기능 및 용도

인류 기술의 발전은 시간이 지날수록 완벽한 인공물을 창작할 수 있다. 오늘날의 디자인은 인간이 사물을 도구로써 충분히 활용할 수 있도록 의미를 부여하는 것이며, 모든 것의 가치는 도구의 사용성에 있다. 디자이너 Otl Aicher는 일상적인 디자인은 '진실의 새로운 기준'[3] 이라며 사용성에 대하여 언급한 바 있다. 이 점에서 디자인의 미학은 예술의 미학과 상당히 다르다. 의도적이지 않고 주변 환경으로 부터 독립적인 가치를 인정받는 예술과는 다르게 디자인은 절대적이고 주관적인 해석이 필요한 미학은 존재하지 않는다. 다시 말해, 디자인은 항상 사물과 인간과의 상호작용에 중점을 둔다.

디자인을 추상적인 맥락에서 이해하고 정당화할 필요는 없다. 사실, 그 반대라고 말할 수 있는데, 디자인은 오로지 만드는 행위를 통해서만 드러난다. 사물의 모습은 더 이상 형태나 미학적 원칙이 아니라 용도에 따라 결정되어야 한다. 형식은 규칙이나 예술에서 나온 것이 아니라 적용을 통한 결과이다.[4] 따라서 사물의 형태는 인간과 상호작용하는 특정한 삶의 방식에서 비롯된다. 사용이란 설명하고자 하는 모든 것을 포기하는 것이다. 사물 그 자체는 그것의 용도로 표현된다.[5] 디자인의 품질은 사용되면서부터 자연스럽게 나타나게 된다. 좋지 않은 디자인은 사용하기 쉽지 않기 때문에 잘못 디자인된 것이다. 반대로 우수한 디자인이라는 것은 특별한 학습 없이 사용하기 편하다. 그렇다면 디자인의 사용품질이라는 것은 어떻게 명확하고 간결하게 정의할 수 있을까라는 질문을 할 수 있다. 어떠한 사용자는 사용하기 쉽다고 생각할 수 있지만 또 다른 사용자가 동일한 평가를 내린다고 말할 수 없다. 이는 사용성을 평가할 때 훨씬 더 구체적인 접근 방식으로 판단해야 함을 의미한다.

미국의 컴퓨터 공학자인 Herbert Simon은 사물의 형태에 대해 단순히 사용성 평가를 넘어선 방법을 제안하고 있다. 그는 사물의 기능성과 목적성에 따른 측면을 인공적인 것과 자연적인 것으로 분류하여 평가한다. 예를 들어, 사람들은 특정한 목적을 위하여 인공물을 생산하는데, Simon은 이러한 인공물의 '인터페이스'를 '내부'환경과 '외부'환경으로 구분하였다. '인공물은 그 자체의 형태와 내/외적인 환경 요소의 접점 (오늘날의 '인터페이스')으로 구성되어 있으며, 둘 사이의 환경이 적절하게 기능할 경우, 인공물의 의도된 목적을 달성한다.'[6] 고 정의하였다. 다시 말해 내부 환경이란, 인공물이 기능하게 하는 요소로 풀이될 수 있으며 일반적으로 도구로서의 목적을 수행하는 것에 맞춰져 있다. 그리고 인공물의 외부환경과도 사용성이 정확히 일치하여야 한다. 이것은 인터렉션의 사용성을 평가하는데 중요한 요소로 볼 수 있다. 예를 들어, 비행기의 기능과 디자인은 인공물을 지상으로부터 날게 하는 데 초점이 맞춰져 있다. 그러나 비행기가 특정 요소에 충족되지 못한다면 제대로 기능할 수 없으며, 관제탑은 사전에 위험요소를 고지하고 안전하게 시스템이 작동할 수 있도록 원활한 지원을 해야만 한다. 또한 공항의 환경은 비행기가 지상에서 쉽게 이동 가능하도록 조성 되어야 한다. 자동차 통행을 목적으로 한 도로와는 다를 수밖에 없을 것이다.

항공기의 운전석은 조종사가 쉽게 조작할 수 있도록 디자인되어야 한다. 모든 디스플레이와 작동 기기가 완벽하게 작동하는 것은 물론, 조종사는 계기판을 확인하거나 조작할 때 안정적인 감정을 느껴야 한다. 하지만 여전히 모든 항공기 관련 재해의 70~80%는 사람의 실수가 원인으로 꼽히고 있으며 비행기는 여전히 복잡한 환경에 대해 완벽히 제어할 수는 없는 요소는 항시 존재한다고 볼 수 있다. 조종사 개개인의 전문성으로 인해 추락의 위험성을 얼마나 예방하였는지는 정확한 통계는 없지만, 비행기를 이용하는 승객들에게는 조종사와 같이 전문적인 기술을 갖고 있는 것은 매우 신뢰적이라고 말할 수 있다. 언젠가는 조종사 없이 비행기가 운용되는 것이 가능하더라도 많은 승객들은 불안한 마음에 조종사가 운전하는 것을 선택할 것이다. 그럼에도 불구하고 우리는 적극적인 기술의 개발로 인해 복잡한 시스템들이 점점 사라지는 것을 볼 수 있다.

≫우리는 적극적인 기술의 개발로 인해 복잡한 시스템들이 점점 사라지는 것을 볼 수 있다≪

인류의 인공 환경은 우리에게 제2의 자연이 되었다. 그리고 기술의 혁신과 끊임없는 발전으로 계속 발전하고 있다. 디자인은 발전하는 산업을 따라가고 있지만 기술의 발전과 혁신의 속도를 따라잡기 어려워지는 것 또한 사실이다. 그럼에도 디자인의 목표는 인간이 최상의 서비스를 누리며 발전되는 기술 환경을 제어하는 것이다. 인간이 한때 기술을 이용하여 원시 자연의 위협으로부터 해방되고 통제했던 것처럼, 이제는 디자인이 기술을 통제하는 역할을 해야 한다. 인공적 자연이 통제되려면 수 많은 제품과 프로세스는 단순화 되어야 하며, 이제는 단순함(Simplicity)은 디자인의 가장 중요한 목표가 되었다. 제품의 생산, 사용자 인터페이스, 시스템, 그리고 작거나 큰 부품 하나까지도 단순함을 추구하는 디자인은 폭넓게 적용되고 있다. 기능의 옵션까지도 줄이는 것이 단순함을 추구하는 효과적인 수단으로 볼 수 있으며 쉽게 조작이 가능하여야 한다. 다시 말해 사용자가 별도의 학습 없이도 바로 사용할 수 있어야 한다.

## 감성

인간은 기술을 통해 자연을 정복했다고 볼 수 있지만, 인간 역시도 자연의 존재이며 귀속된 부분이다. 인류의 발명품은 자연환경을 이용하거나 극복하기 위한 목적으로 만들어졌기 때문에 자연환경의 법칙에 대하여 소통하는 것이라고 말할 수 있다. 새로운 인간의 기술은 이같이 끊임없이 자연에게 영향을 주고 변화되고 있다. 인간의 목적이 바뀌면 인공물도 바뀌게 된다. 혹은 그 반대의 순서도 가능하다.[7] 각각의 목적은 동기를 기반으로 하고 있으며, 이는 모두 감성에 기반을 두고 있다. 생물학자 Humberto Maturana는 감성을 인간 본성의 근본적인 조건으로 설명하고 있다. "감성은 일반적인 동물, 특히 우리 인간이 모든 순간을 움직이게 하는 원동력이라고 말할 수 있다."[8]

≫감성은 일반적인 동물, 특히 우리 인간이 모든 순간을 움직이게 하는 원동력이라고 말할 수 있다.≪

이것은 모든 인간의 행동, 그리고 목적을 설정하는 것이 감성에 기반을 두고 있음을 의미한다. 하지만 인간과는 달리 기계나 시스템을 수행하는 인공물은 감성에 의해 기능하지 않는다. 인공지능 역시도 감성에 의하여 스스로 목적을 설정하거나 행동할 수 없다. 오늘날의 관점에서는 기계가 이러한 자연스러운 단계에 도달한다는 것은 상상할 수 없다.

그렇다면 이러한 측면은 로봇이나 지능형 무인 운송 시스템, 항공기 등과 같은 미래의 인공물 디자인에 무엇을 의미할까? 후에 감성으로 움직이게 되는 유토피아적 혁신이 가능하다 하더라도, 이것을 움직이게 하는 것은 여전히 인간일 것이다. 이 같은 점에서 디자이너들은 감성과 기계의 연결고리를 만들어내는 역할을 하게 된다. 자연에서 혁신적인 영감을 얻거나 스타일링을 하게 되며, 가장 중요한 감성적인 요소를 부여하기 위해 노력한다.

인공물에는 감정이 전혀 없지만 인간의 행동은 상당 부분 감성에 의해 결정된다고 말할 수 있다. 이것은 인공물과의 상호작용에 대해서도 마찬가지인데, 냉장고, 노트북, 로봇을 조종하든, 사람의 행동에는 언제나 감정이 따른다. 주변에 있는 인공물들이 사용자 감정에 대해 알지 못하지만, 사용자는 인공물에게 기쁨, 동정, 무관심, 심지어는 불안함이나 두려움을 느낄 수도 있다. 이것은 디자이너들이 긍정적 감성이 느껴지게 하는 디자인을 하게 되는 주된 이유이기도 하다. 이러한 예는 우리가 매일 사용하게 되는 사물에 적용되어 있는 것을 알 수 있는데, 예를 들어 개인용 로봇은 사용자와 감정적 교감을 불러일으키도록 유아적인 감성으로 디자인된다. 그리고 제품에 대해 어린아이와 같은 특성을 부여하게 되면 무해하게 보여지며 사용자로 하여금 조작하는 것에 대한 두려움이 발생되지 않는다. 또한 감정이 없는 인공물이 살아있는 존재로 인식된다.

비행기는 로봇 새가 아니다. 완전히 다른 방식으로 작동하지만 날 수 있다. 마찬가지로 로봇은 인간과 다르게 작동하지만 인간을 위해 수많은 일을 할 수 있다. 1920년에 체코 작가 Karel Čapek의 드라마 'R.U.R. - Rossum's Universal Robots'에서는 만능 로봇이 인류에게 종말을 예고하는 것으로 화제가 되었다: "세계의 로봇이여! 많은 사람들이 타락했다. 우리는 공장을 인수했고 이제 세계의 주인이다. 인간의 시대는 끝났다. 새로운 시대가 도래했다! 로봇이 지배하는 세상!"[9] 의 대사가 방송된 바 있다. 당시에는 2차 산업혁명이 한창이던 때였고, 생산 라인과 전기의 도입으로 공장 자동화로 대량생산이 시작되는 시대였다. 수많은 비평가들은 새로운 노동조건이 비인간적이라며 비난했다. 생산 라인 작업자는 자동화된 프로세스의 일부가 되었으며 작업이 몇 번의 단조로운 손 동작으로 축소되었다. 이러한 시대적 상황은 수많은 책, 연극 및 영화에서도 강조되고 비판되었다. 따라서 Čapek의 드라마는 이러한 노동조건에 대한 반응으로 볼 수도 있다. 생산 라인의 단조롭고 간단한 작업은 인공 작업자(로봇)가 인간을 대체할 것이라는 개념을 형성하는 데 중요한 역할을 하게 되었다.

## 자연적인 것과 인위적인 것

유토피아적 환상은 현실이 되었다. 서두에서 언급했듯 대량 생산 기반의 산업은 디자인의 중요성과 함께 성장하게 되었다. 선구자적인 디자이너들은 심미성만을 표현하는 화장품과 같은 표면적 가치를 넘어서야 한다는 것을 인식했고, 제품 전체의 개발 프로세스에 영향을 갖고자 노력하였다. 오늘날에는 당연하게 되었지만 당시 많은 기업가들은 디자인과 산업적인 융합에 대하여 많은 의구심이 존재하였지만, 최상의 해답을 위하여 디자이너와 엔지니어는 끊임없이 협업으로 작업하게 되었다. 그렇다면 미래의 제품은 상상과 AI가 설계하고 개발하게 된다면, 디자인은 어떤 역할을 해야 할까?

미국 저널리스트 Kevin Kelly는 자신의 저서 'Out of Control'에서 기계는 생물학적으로 변하고 생물학적인 것은 공학적으로 변하고 있다고 한다. 그는 이러한 현상을 '자연물과 인공물의 결합'[10]이라고 설명한다.

≫자연물과 인공물이 결합하여 새로운 '유기적 인공물' 세상을 만들 것이다≪

자연물과 인공물이 결합하여 새로운 '유기적 인공물' 세상을 만들 것이다. 이러한 발전은 인간의 새로운 환경을 제공할 것이다. 또한 인간은 다시 한번 새로운 기술 환경에 대한 통제권을 상실하는 것을 의미할 것이다. 인류가 만든 세계는 이미 너무 복잡해져서 이를 관리하기 위해서는 통제 가능했던 환경을 기억해야 한다. 즉, 우리를 구성하고 있는 환경이 기계적으로 될수록 유기적 환경을 조성하여야 한다.[11] 그러므로 인류는 새로운 과학적 미래로 향해 나아갈 수 있으며, Kelly는 이를 '신생물학적 문명'이라고 부른다. 이처럼 반대되는 두 환경 요소에 대하여 Kelly는 다음과 같이 설명하고 있다. "인간이 만든 것은 유기적으로 되어가고 있으며, 생물학적 요소는 기계적으로 되어간다."[12]

사물들이 지금보다 유기적인 존재가 된다면 어떤 결과가 나타날까? 첫째로는 스스로 존재를 정립하고 진화하게 될 것이다. 그러기 위해서는 새로운 방법으로 기능해야 할 것이다. 이러한 기술들은 서로 소통하고, 인식하며, 지식의 발전을 거듭해야 한다. 그렇다고 새로운 인공물들이 기존 생명체에 필적하고 위협한다는 의미는 아니다. 왜냐하면 새로운 인공물들은 감정에 기반하여 기능하는 것이 아니라, '조력자'로서 존재하기 때문이다. 기본적인 조력자로서의 개념은 이메일 보조 서비스로 이해할 수 있다. 조직 간 토론이나 소통을 쉽게 보조할 수 있는 소프트웨어의 개념이다. 이러한 사례는 인공지능의 한 가지 예로 볼 수 있으며, 언어 분석과 작업 처리 기능으로 방대한 양의 업무를 수행하는 것으로 사용하고 있다.[13]

## 호모 엑스 데이터 (데이터 기반 인간)

그렇다면 인간의 삶이 더욱 공학적으로 변해간다는 것은 무엇을 의미할까? Kevin Kelly는 이를 인간이 기술이 '통제할 수 없는' 상태라고 정의한다. 인류를 구성하고 있는 환경은 변화하고 있으며, 사물들이 서로 '소통'하고 '교감'하는 것과 같이 인간/인공물은 동등하게 인공지능에 적용되고 있다. 더 이상 객체별 제품이나 사물을 디자인하는 것이 아니라 새로운 생활환경 안에 구성 요소를 창조하는 것이 되었다.

동시에 인류는 더욱더 사물을 간소화하고 쉽게 바꿔나가고 있다. 과학과 인간은 협업하며 자연과 인공물을 바꿔나가고 있으며, 주변 환경은 셀 수 없을 정도의 수많은 센서와 데이터를 등록하는 네트워크가 되어 새로운 환경을 생성하게 된다. 이러한 신세계는 진정한 의미의 지속적 최적화 과정이라 볼 수 있다. 소비자로써 사용자는 제품의 새로운 기능을 습득하게 되고, 반대로 제품은 사용자로부터 새로운 기능을 형성할 수 있다. 이러한 구성 환경은 지속적인 센서 제어와 데이터 전송의 구조로 가능하게 된다.

이런 기술들은 필연적으로 인류 발전에 커다란 영향을 미칠 것이다. 호모 사피엔스와 호모 파베르에 이어 새로운 유형의 호모 엑스 데이터라는 새로운 인류가 등장하게 될 것이다. 즉, 더 이상의 생물학적 전통이나 기술을 이용하는 환경에 의존하지 않으며, 데이터 생성과 전송에 의하여 생활 환경이 결정되는 인류가 될 것이다. 호모 엑스 데이터 인류는 더 이상 물리적 환경을 변화시키지 않을 것이며, 점점 더 데이터의 규모와 수집, 평가에 의존하게 될 것이다.

≫호모 엑스 데이터 인류는 더 이상 물리적 환경을 변화시키지 않을 것이며, 점점 더 데이터의 규모와 수집, 평가에 의존하게 될 것이다≪

기능성과 경험을 바탕으로 제품을 개발하는 일은 이제 드문 경우가 되고 있으며, 특정 데이터의 수집과 분석으로 최적화 프로세스를 통한 제품 개발로 변해가고 있다. 제품 디자인도 이러한 데이터 기반 산업에서 벗어날 수 없는데, 바로 사용자 경험에 의한 데이터 축적이다. 극단적으로는 데이터 기반의 디자인을 벗어나는 것은 심미적 용도의 액세서리 디자인에 한정될 것이며, 대부분의 호모 엑스 데이터 인류는 시스템적으로 사물을 결정할 것이다. 모든 사물의 기능, 목적은 센서를 기준으로 결정되고 제작될 것이다. 사물의 오류 발생 데이터는 즉시 네트워크를 통하여 '전달' 될 것이며, 업데이트되어 동기화될 것이다. 디자인 다양성이라는 특성은 소멸될 확률도 크다고 볼 수 있다.

일상생활의 사물의 제작도 데이터화되고 있다. 이러한 변화는 호모 엑스 데이터 인류가 직접 제품 설계 및 제작에 참여할 수 있게 되는데, 참여를 통한 데이터 수집으로 제품의 발전하게 되는 계기를 제공한다. 제품들은 다른 제품들과 정보가 수집, 분석되어 네트워크화된다. 이것은 유비쿼터스 센서 시스템에 의한 감시, 관찰, 평가되어 데이터 투명성과 분석으로 생활에 반영되게 된다. 이것은 일반적으로 '빅 데이터' 라는 용어로 설명되는 현상이다. 측정할 수 없을 정도로 광범위한 데이터 소스를 사용하여 현재의 현상에 대해 믿을 수 없을 정도의 빠른 수집, 분석, 평가되는 데이터를 말한다. 자연, 인공적 환경, 생물 어디에나 센서가 측정될 수 있다면 모든 것들이 데이터 소스로 사용될 수 있다.
영화로도 제작된 미국 작가 Dave Eggers의 베스트셀러 소설 'The Circle' 은 사회가 완전한 투명성, 감시, 데이터 통제에 의해 지배되는 무서운 미래를 그리고 있다.
호모 엑스 데이터와 함께 인간의 진화는 완전히 새로운 단계로 접어들고 있다. 호모 파베르와 달리 이 새로운 유형의 인류는 더 이상 환경을 통제하는 기술에 대한 권력을 쟁취하는 데 관심이 없다. 그 대신 호모엑스 데이터 인류는 완벽하고 투명한 데이터를 구성하고 싶어 한다. 그리고 자연과 인공물로 거듭난 유기적 인공물의 환경은 나날이 새롭게 구성되고 있다. 센서로 둘러싸인 새로운 인류는 스스로 조직적인 구성과 갱신으로 세상을 만들어가고 있다. 이 점이 호모 엑스 데이터 인류의 생활은 호모 파베르와 다르다고 볼 수 있다. 독립적으로 스스로를 평가하고 경험할 뿐 만 아니라, 소유에 대한 개념은 보편적인 자아실현을 위해 덜 중요해진다.

이러한 다양한 서비스들은 삶의 질을 경험하는데 개별 제품 디자인들보다 커다란 영향을 끼친다. 기능성은  제품 자체의 가치보다 맥락이 중요하게 평가되며, 중요성은 데이터의 일부가 네트워크로 전송되어 평가받게 된다.

≫호모 엑스 데이터와 함께 인간의 진화는 완전히 새로운 단계에 들어서고 있다. 호모 파베르와 달리 이 새로운 유형의 인류는 완벽하고 투명한 데이터를 구성하고 싶어 한다. 그리고 자연과 인공물로 거듭난 유기적 인공물의 환경은 나날이 새롭게 구성되고 있다≪

동시에 생산과 소유의 개념은 끊임없이 변화하게 되는데, 기계적이고 물리적인 생산과 개념에서 디지털과 같은 무형의 제어시스템으로 대체되기 때문이다. 이것은 다양하고 유연한 세상을 가능하게 한다. '소통'과 '상호작용'은 인간과 시스템의 공존을 가능하게 하며, 극히 개인적인 제품을 위한 디자인보다는 보편적이고 사회적인 디자인이 지배하게 된다.

호모 엑스 데이터 인류와 마찬가지로 새로운 제품의 엑스 데이터는 맥락이라는 요소가 더욱 중요하게 된다. 이것은 새로운 디자인의 도전 과제라고 할 수 있으며, 데이터의 세상에서 더 이상의 개별적인 제품 개발은 존재하지 않는다는 것을 의미한다. Dave Eggers가 그의 소설에서 묘사한 인간과 인공물의 상호작용에 대해서 '행동의 영역은 감성에 의해 기능한다'[15]고 한 바와 같이, 자연과 인공물이 점점 더 밀접하게 연결되고 있지만, 아직까지는 감정적 동기를 만들어낼 수 있는 것은 인간뿐이라 말할 수 있을 것이다.

[1]  Gert Selle, "Designgeschichte in Deutschland. Produktkultur als Entwurf und Erfahrung" (Cologne, 1987), p. 20.
[2]  Gert Selle: loc. cit., p. 7.
[3]  Otl Aicher, "Analogous and Digital" (Hoboken, NJ, 2015), p. 88.
[4]  Otl Aicher: loc. cit., p. 91.
[5]  Otl Aicher: loc. cit., p. 92.
[6]  Herbert A. Simon, "The Sciences of the Artificial" (Cambridge, 1969), p. 6.
[7]  Herbert A. Simon: loc. cit., p. 3.
[8]  Humberto R. Maturana, "Biologie der Realität" (Frankfurt, 1998), p. 325.
[9]  Quoted from Red Dot 21. www.red-dot-21.com.
[10]  Kevin Kelly, "Out of Control: The Rise of Neo-Biological Civilization" (New York, 1994), p. 2.
[11]  Kevin Kelly: loc. cit., p. 2.
[12]  Kevin Kelly: loc. cit., p. 3.
[13]  Francisco J. Varela, "Kognitionswissenschaft – Kognitionstechnik" (Frankfurt, 1990), p. 114.
[14]  Dave Eggers, "The Circle" (New York, 2013).
[15]  Humberto R. Maturana: loc. cit.

»Following *Homo sapiens* and *Homo faber*, a new type of human, which we call *Homo ex data*, is emerging. Fitted with and surrounded by sensors, *Homo ex data* becomes part of a systematically constructed 'bio-artificial' reality.«

Peter Zec

≫호모 사피엔스, 호모 파베르에 이어 호모 엑스 데이터라는 새로운 인류가 등장하고 있다. 센서로 둘러싸인 호모 엑스 데이터 인류는 '유기적 인공물'로 연결된 환경의 일부로써 귀속될 것이다.≪

피터 젝

# Activities and Products in the World of Homo Ex Data

# 세계 속의
# 호모 엑스 데이터 사례

On the following pages, this book introduces products that make *Homo ex data* possible in the first place. They include products that help collect, process and use data. All of these products have been successful in the Red Dot Design Award.

이 책에서는 호모 엑스 *데이터*가 기능하는 제품들을 소개한다. 소개된 제품들 모두 데이터를 활용하는 제품이며, 수록된 사례 모두 레드닷 디자인 어워드에서 수상되었다.

# Monitoring 모니터링

The collection of data on human beings often already begins before they are born – probably in the course of ultra-sound examinations and prenatal diagnostics. Data on foetuses, including, in part, even genetic profiles, are collected, analysed, or compared with reference data in order to create an initial prognosis and diagnosis for the as yet unborn life. Today, human beings are, therefore, already part of the data system often even before they are born – and, thus, digital natives in the true meaning of the word, original natives of the digital world. From the very beginning, they become data-based humans – *Homo ex data*.

Despite all the possible existing ethical concerns with respect to such early data analysis and decisions derived from it, there are many medical cases in which monitoring human data makes sense, if it is not even necessary for survival. If a child comes into the world long before its anticipated due date, it is not yet able to survive on its own. In order to survive, it depends on technology – as well as monitoring and evaluation of its body data. Incubators such

as the Dräger Babyleo TN500 IncuWarmer provide premature babies with an optimal microclimate. Highly sensitive sensors facilitate constant monitoring of vital functions as well as of temperature, humidity, oxygen content, and brightness and noise level inside the incubator, and hence reduce work for medical personnel. The still fragile life depends on a stream of data, while the continuous transfer of data links the natural with the artificial, thus assuring survival. This aspect is also expressed in the design of the Babyleo: transparent sides allow viewing and provide closeness to the baby; at the same time the baby lies in a kind of artificial, engineered womb, which screens and protects the child from the surroundings. Thanks to its sensitive design, the Babyleo succeeds in instilling confidence in the basic technology and creating a calming atmosphere.

인간에 대한 데이터 수집은 보통 태어나기 전부터 이미 시작된다. 아마도 초음파 검사와 산전 진단 과정에서 수집되기 시작할 것이다. 아직 태어나지 않은 생명에 대한 초기 예측과 진단을 하기 위해 부분적으로 유전적 특성을 포함한 태아에 대한 데이터를 수집, 분석, 또는 참조 데이터와 비교한다. 따라서 오늘날 인간은 흔히 태어나기도 전에 이미 데이터 시스템의 일부가 되며, 진정한 의미에서 디지털 원주민, 즉 디지털 세계의 원주민이 된다. 맨 처음부터 그들은 데이터 기반 인간인 호모 엑스 데이터(Homo ex data)가 된다.

이러한 초기 데이터 분석 및 그로부터 파생된 결정과 관련하여 기존의 가능한 모든 윤리적 문제에도 불구하고, 생존을 위해 필요하지 않은 경우에도 인간 데이터의 감시가 의미가 있었던 의학적 사례가 많이 있다. 태아는 예정된 기한보다 일찍 태어나게 되면 혼자 살아남을 수 없다. 살

아남기 위해서는 신체 데이터를 감시하고 평가하는 것은 물론 기술에 의존해야 한다. Dräger Babyleo TN500 IncuWarmer와 같은 인큐베이터는 미숙아에게 최적의 환경을 제공한다. 고감도 센서는 인큐베이터 내부의 온도, 습도, 산소 함량, 밝기 및 소음 수준은 물론 중요한 신체기능을 지속적으로 용이하게 감시하여 의료인력의 부담을 줄여준다. 여전히 연약한 생명은 데이터의 흐름에 의존하는 반면, 데이터의 지속적인 전송은 자연과 인공을 연결하여 생존을 보장한다. 이것은 Babyleo의 디자인에서도 표현된다. 투명한 쪽은 아기를 볼 수 있게 해주고 아기에게 친밀감을 제공한다. 동시에 아기를 일종의 인공 자궁에 위치시켜 주변 환경으로부터 아이를 보호한다. Babyleo는 이러한 근본적이고 중요한 환경을 조성하는데 성공했다.

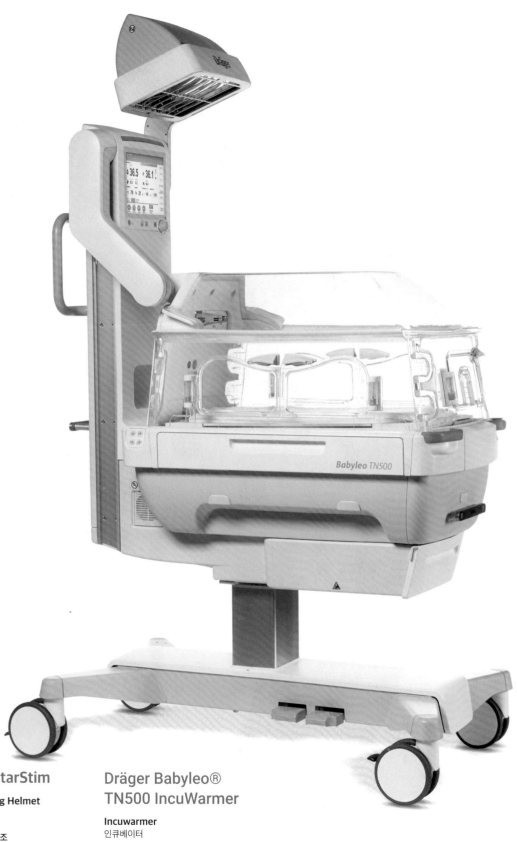

## Enobio 2 StarStim

**Brain-Monitoring Helmet**
뇌 모니터링 헬멧

**Manufacturer 제조**
Neuroelectrics
Barcelona | Spain
**Design 디자인**
ÀNIMA design, ÀNIMA Barcelona
바르셀로나 | 스페인
www.neuroelectrics.com
Page 34 | 35

## Dräger Babyleo®
## TN500 IncuWarmer

**Incuwarmer**
인큐베이터

**Manufacturer 제조**
Drägerwerk AG & Co. KGaA
Lübeck | Germany
**Design 디자인**
MMID GmbH
에센 | 독일
www.draeger.com

acking

추적

## Self-Tracking 자가추적

Self-monitoring or self-tracking is a worldwide trend. Today, about two billion people around the world already monitor their bodies by means of wearables and apps. In the process, vast quantities of personal data are collected and stored and become part of Big Data. From a technological perspective, the trend towards self-tracking is fuelled by the development of biometric sensors that are becoming ever smaller and more cost-effective and by today's ubiquitous smartphones as well as the possibility of storing and processing ever greater amounts of data ever more quickly. It has therefore become easy to monitor or measure aspects such as heartbeat and blood pressure, movement when awake or asleep, menstruation cycle, eating habits, changes in weight and similar matters, and to record and evaluate them. All of this is conducive not only to enhanced self-awareness, but ultimately also to systematic self-optimisation. In spite of more precise sensors for monitoring bodily functions, tracking fitness and health is less a medical issue than a question of lifestyle. Accordingly, portable measuring instruments to be worn on the body, so-called wearables, are also designed more as decoration or lifestyle accessories than as health items.

One prime example of this is the Motiv Ring. It is a micro-fitness tracker able to measure activity, pulse frequency, and sleep quality that is worn on the finger. Sensors, battery, and Bluetooth receiver are accommodated in the smallest space possible. The majority of fitness and health trackers are, however, inspired by classical watch design – thus, for example, the Apple Watch Series 2: the smartwatch is equipped with a wide range of fitness and health functions, while integrated GPS ensures that speed and distances covered can be measured, even without an iPhone.

자가 모니터링 또는 자가추적은 세계적인 추세이다. 오늘날 전 세계적으로 약 20억 명의 사람들이 이미 웨어러블과 앱을 통해 자신의 신체를 모니터링하고 있다. 이 과정에서 방대한 양의 개인 데이터가 수집 및 저장되고 빅 데이터의 일부가 된다. 기술적인 관점에서 볼 때, 자가추적에 대한 추세는, 점점 크기가 더 작아지면서도 비용 효율성이 높아지는 생체인식 센서의 개발과 오늘날의 유비쿼터스 스마트폰, 그리고 훨씬 더 많은 양의 데이터를 저장하고 처리할 수 있는 가능성에 의해 촉진된다. 따라서 심장 박동 및 혈압, 기상 또는 수면 시 움직임, 월경 주기, 식습관, 체중 변화 및 이와 유사한 사항과 같은 측면을 모니터링 또는 측정하고, 이를 기록하고 평가하는 것이 쉬워졌다. 이 모든 것이 자기 인식 향상에 도움이 될 뿐만 아니라 궁극적으로 체계적인 자기 최적화에도 도움이 된다. 신체 기능을 모니터링하기 위한 센서가 더욱 정밀하지만, 신체 단련 및 건강을 추적하는 것은 의학적 문제가 아니라 생활방식의 문제다. 신체에 착용하는 휴대용 측정기, 이른바 웨어러블도 건강용품이기보다는 장식이나 생활용품으로 디자인되고 있다.

이에 대한 한 가지 대표적인 예가 Motiv Ring이다. 손가락에 착용하는 활동량, 맥박수, 수면의 질을 측정할 수 있는 초소형 피트니스 추적기이다. 센서, 배터리, 블루투스 수신기는 가능한 가장 작은 공간에 수용된다. 그러나 대부분의 피트니스 및 건강 추적기는 고전적인 시계 디자인에서 영감을 얻었다. 예를 들어 Apple Watch Series 2: smartwatch에는 광범위한 피트니스 및 건강 기능이 장착되어 있으며 통합 GPS는 속도와 거리를 보장한다. 아이폰 없이도 이동한 거리를 측정할 수 있다.

## Motiv Ring

**Fitness Tracker**
피트니스 트랙커

**Manufacturer 제조**
Motiv Inc.
샌프란시스코 | 캘리포니아 | 미국
**In-house design 인하우스 디자인**
Curt von Badinski
www.mymotiv.com

1

4

# 1
## Amazfit

**Activity Tracker**
액티비티 트랙커

**Manufacturer 제조**
Huami Inc
마운틴 뷰 | 캘리포니아 | 미국
**In-house design 인하우스 디자인**
www.amazfit.com

# 2
## Apple Watch Series 2

**Smartwatch**
스마트 워치

**Manufacturer 제조**
Apple
쿠퍼티노 | 캘리포니아 | 미국
**In-house design 인하우스 디자인**
www.apple.com

# 3
## MOOV NOW™

**Sports Tracker**
스포츠 트랙커

**Manufacturer 제조**
Moov
벌링게임 | 캘리포니아 | 미국
**In-house design 인하우스 디자인**
www.moov.cc

# 4
## vivomove Asia Edition

**Fitness Tracker**
피트니스 트랙커

**Manufacturer 제조**
Garmin
신베이 | 타이완
**In-house design 인하우스 디자인**
Chun-Kai Huang | Han-Wei Huang
www.garmin.com.hk

**2**

**3**

**5**

# 5

## Mio Slice Band

**Heart Rate and Health Tracker**
심박수 건강 트랙커

**Manufacturer 제조**
Mio Global
밴쿠버 | 캐나다
**Design 디자인**
Woke Studio Inc
밴쿠버 | 캐나다
www.mioglobal.com

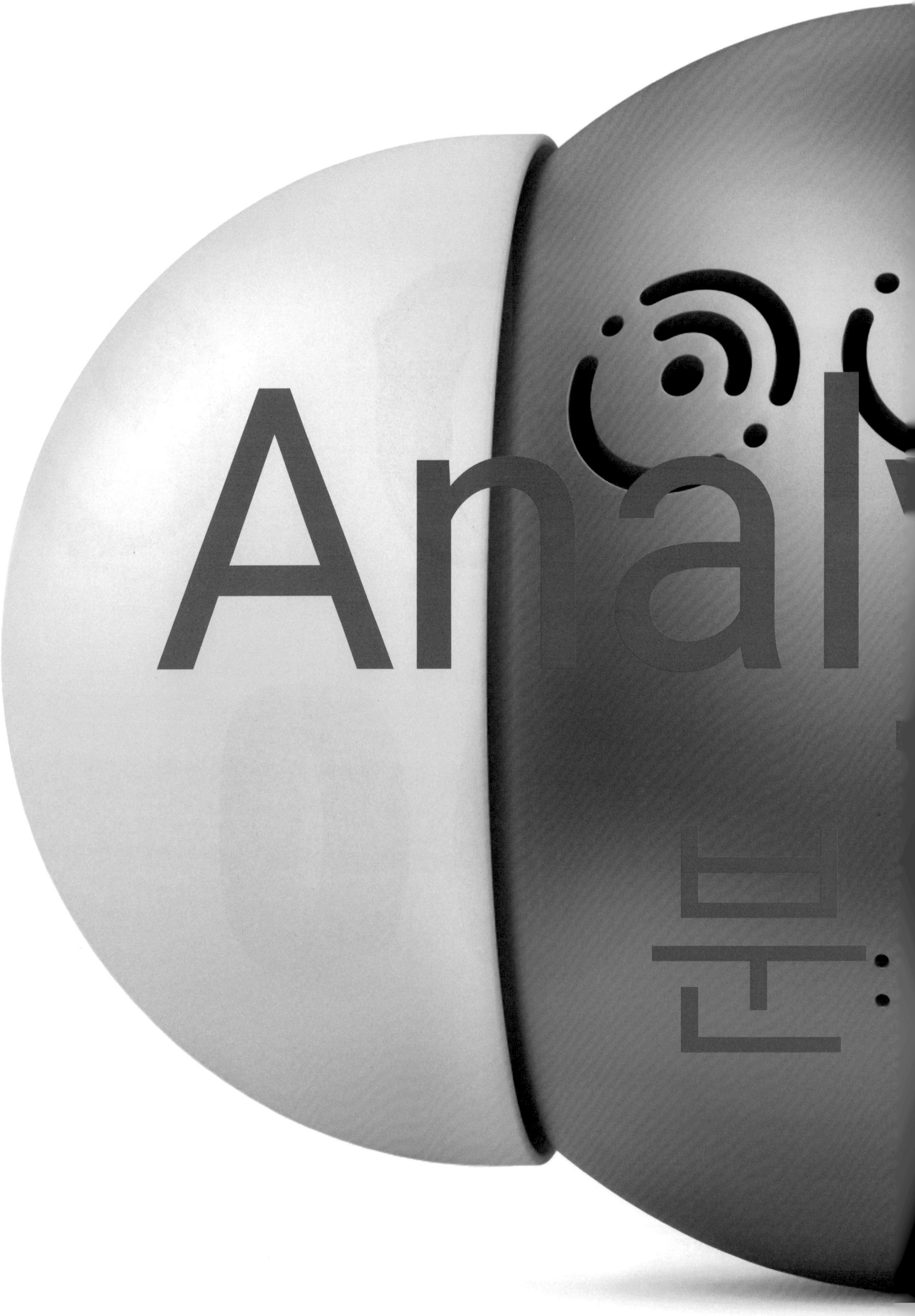
Anal

## Analysing 분석

Not only bodily functions have been monitored since time immemorial. The environment with its weather phenomena such as drought, rainy periods, storms, etc. has also shaped people's lives and, therefore, been observed for thousands of years. What is relatively new, in contrast, is the possibility to conduct in-depth analysis not only of the macro-, meso- and microclimate, but also of one's very personal surrounding environment. It is therefore about ecology in the original meaning of the term, according to which this discipline examines the reciprocal relationship between living creatures and their environment. Thus, ever more frequently, wearables collect not only body data, but, thanks to corresponding sensors, also measurable data from the environment. Through the measuring of pollutants and fine dust in the air, ozone values and UV radiation, temperature, humidity, loud noises, and radioactivity, the "Quantified Self" is then coupled with the "Quantified Environment". However, particularly with respect to the measurement of the environment, products that are used exclusively for this purpose still promise greater precision – simply because the requirement that wearables be of the smallest possible size and thickness make them (as yet) unable to accommodate very many components. One example of a mobile air sensor that – similar to a fitness tracker – is designed like a fashionable accessory is the Air Dragon. This small device is equipped with an air quality sensor that detects pollutants, for example, exhaled breath, nicotine, solvents, or paint in the ambient air, and warns the user if limit values are exceeded.

옛날부터 신체 기능만 관찰된 것은 아니다. 가뭄, 장마, 폭풍 등과 같은 기상 현상이 있는 환경도 사람들의 삶에 영향을 주어 왔으며, 이것은 수천 년 동안 관찰되었다. 현대에 와서는 사용자의 아주 가까운 환경을 분석할 수 있는 가능성이 열리게 되었고, 이 분야는 생물과 환경 사이의 상호 관계를 조사하는 기능을 수행하게 된다. 따라서 웨어러블은 신체 데이터는 물론 해당 센서를 통해 환경에서 측정 가능한 데이터도 수집하는 경우가 더 많아지고 있다. 공기 중의 오염물질과 미세먼지, 오존값과 자외선, 온도, 습도, 큰 소음, 방사능 등을 측정하여 '사용자의 상태'와 '환경 상태'를 연결한다. 특히 환경 측정과 관련하여 이 목적으로만 사용되는 제품은 여전히 더 높은 정확도를 보장한다. 웨어러블이 가능한 한 가장 작아야 하고 두께가 얇아야 한다는 요구사항으로 인해 (아직) 많은 구성요소를 수용할 수 없기 때문이다. 피트니스 추적기와 유사하게 패셔너블한 액세서리처럼 디자인된 모바일 공기 센서의 한 예는 Air Dragon 이다. 이 소형 장치에는 대기 중 오염 물질(예: 호기, 니코틴, 솔벤트 또는 페인트)을 감지하고 한계값을 초과하면 사용자에게 경고하는 공기 품질 센서가 장착되어 있다.

자동차, 주택, 사무실 건물, 작업장 또는 도시 전체에서 환경 데이터를 측정할 때 무엇보다도 정밀도가 높은 센서가 필수적이다. 이 경우 환경적

# Air Dragon

**Air Monitoring Device**
공기 모니터링 장치

**Manufacturer 제조**
Beijing Coilabs Co., Ltd.
베이징 | 중국
**Design 디자인**
LKK Design Beijing Co., Ltd.
베이징 | 중국

When measuring environmental data in cars, houses, office buildings, workshops, or even whole towns, highly precise sensors are above all essential. In this case, it is necessary to analyse and evaluate environmental values quickly in order to, if needed, react to dangerously high values and be able to initiate countermeasures. With its high measuring sensitivity, the Smart Carbon Monoxide Sensor is one such product for networked living environments. Even if there are only very low carbon monoxide emissions, a text on the smartphone warns the occupant and the sensor also activates the ventilation systems and fans that are respectively linked with it.

The Soil Scanner, a scanner developed in the Netherlands that is designed for use in developing countries, provides a quite different kind of environmental analysis. Data gathered by the Soil Scanner are transmitted to a smartphone app and analysed. Farmers then receive recommendations for how they can optimally plant or fertilise their farmland. The entire process – from scanning the soil, to analysis, to providing recommendations – takes less than ten minutes.

가치를 빠르게 분석하고 평가하여 필요한 경우 위험하게 높게 측정된 값에 대응하고 대응책을 적용시킬 수 있어야 한다. 측정 감도가 높은 스마트 일산화탄소 센서는 네트워크 생활환경에 적합한 제품 중 하나다. 아주 적은 양의 일산화탄소만 배출되더라도 스마트폰의 문자가 탑승자에게 경고하고 센서는 각각 연결된 환기 시스템과 팬을 작동시킨다.

개발 도상국에서 사용하도록 네덜란드에서 디자인, 개발된 스캐너인 Soil Scanner는 완전히 다른 종류의 환경 분석을 제공한다. 토양 스캐너에서 수집된 데이터는 스마트폰 앱으로 전송되어 분석된다. 그런 다음 농부들은 농지를 최적으로 심거나 비옥하게 할 수 있는 방법에 대한 권장 사항을 전달받는다. 토양 스캔에서 분석, 권장사항 제공에 이르기까지 전체 프로세스는 10분 미만이 소요된다.

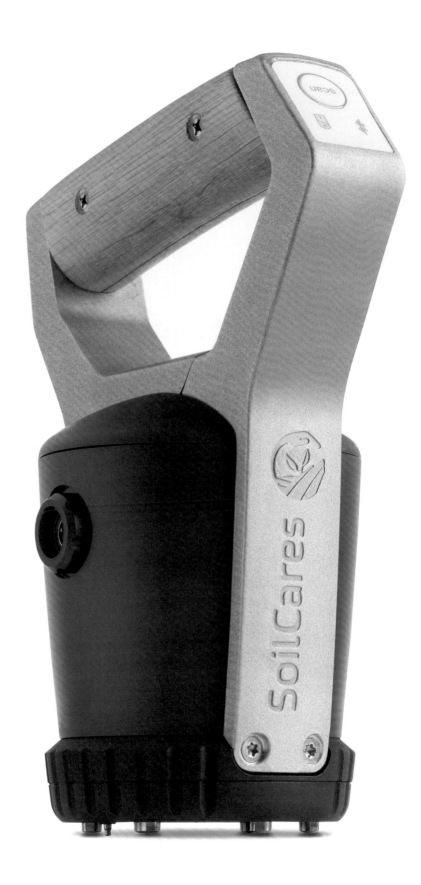

# Soil Scanner

토양 스캐너

**Manufacturer 제조**
SoilCares
바헤닝언 | 네덜란드
**Design 디자인**
Scope Design & Strategy bv
(Pim Jonkman)
아메르스포르트 | 네덜란드
www.soilcares.com

**1**

**2**

**4**

**1**

### Smart Carbon
### Monoxide Sensor

**CO-Detector**
일산화탄소 감지기

**Manufacturer 제조**
Shenzhen Heiman
Technology Co., Ltd.
선전 | 중국
**In-house design 인하우스 디자인**
Zeng Yonggang, Shu Pan,
Peng Shoukun
www.heimantech.com

**2**

### GIS 1000 C
### Professional

**Thermo Detector**
열 감지기

**Manufacturer 제조**
Robert Bosch
Elektrowerkzeuge GmbH
라인펠덴에히터딩엔 | 독일
**Design 디자인**
TEAMS Design
에슬링겐 | 독일
www.bosch-pt.com

**3**

### MyAir

**Portable Air Sensor**
휴대용 공기 센서

**Manufacturer 제조**
Tion LLC
노보시비르스크 | 러시아
**Design 디자인**
Logeeks LLC
노보시비르스크 | 러시아
www.tion.ru

**4**

### Cigsor

**Humidor Sensor**
습도 센서

**Manufacturer 제조**
Cigsor AG
테게르빌렌 | 스위스
**In-house design 인하우스 디자인**
www.cigsor.com

3

5

## 5
## AirVibe

**Air Quality Sensor**
공기 품질 센서

**Manufacturer 제조**
Philips
에인트호번 | 네덜란드
**In-house design 인하우스 디자인**
Philips Design
에인트호번 | 네덜란드
www.philips.com

# Measuring 측정

Measuring the world is a necessity for humankind. For at least 4,000 years, measurements have helped define property boundaries and thus secure one's property; they make the world more manageable, facilitate better travel planning, and hence extend each individual's sphere of action. Whereby, in the past, hemp ropes segmented by means of knots were used to measure distances and define them on maps, today GPS satellites, aerial photographs, and laser scanners have simplified the recording and measuring of distances, surface areas, and spaces. We live in a digitally mapped world, and the omnipresence of such data makes it possible for us for us to access them at any time. We can, therefore, be shown exactly the section of a map that we currently need in real time so as to navigate, plan, or explore it digitally. On the other hand, we can easily collect cartographic data ourselves today. Because, particularly as a result of laser scanning, capturing spaces digitally has become so easy that people are today able to measure their own surroundings. The corresponding devices facilitate cost-effective and, above all, very rapid, three-dimensional measurements of buildings. It is therefore possible not only to capture ground plans, surface areas, and volumes, but also to generate, in part, quite detailed 3D views.

The design concepts underlying these laser scanners differ, in part, significantly: the optical CMM 3D scanner MetraSCAN 3D therefore has a modular design in order to take into account the requirements of engineers, who under professional conditions require very exact measurement results. With this system, highly complex data can be collected with absolute precision. The development and design of the image-producing laser scanner Leica BLK360, on the other hand, were guided by the premise of making the sophisticated measuring technology available to everyone. It transposes the complex underlying technology in an extremely reduced form. This scanner is controlled by one single button. Once it has been activated, the scanner rotates and records its surroundings. An app filters and registers the scan data in real time. Based on these data, users are supplied with full colour, high-resolution, 3D panorama images in the simplest way possible.

세계를 측량하는 것은 인류에게 불가피한 일이다. 최소 4,000년 동안 측량은 재산의 경계를 정의하여 재산을 보호하는 데 도움이 되었다. 측량은 세상을 더 관리하기 쉽게 만들고 더 나은 이동 계획을 용이하게 하여 각 개인의 행동 범위를 확장했다. 과거에는 매듭으로 분할된 대마 로프를 사용하여 거리를 측정하고 지도에 정의했지만, 오늘날에는 GPS 위성, 항공 사진 및 레이저 스캐너를 이용하여 거리, 표면적, 공간의 기록과 측량을 단순화한다. 우리는 디지털로 매핑된 세계에 살고 있으며 그러한 데이터의 편재로 인해 언제든지 이를 액세스할 수 있다. 그래서 현재 필요한 지도 구역을 실시간으로 정확하게 볼 수 있어, 디지털 방식으로 길을 찾고, 계획하거나, 탐색할 수 있다. 한편 오늘날에는 쉽게 지도 제작 데이터를 스스로 수집할 수 있다. 특히 레이저 스캐닝의 결과로 디지털 방식으로 공간을 캡처하는 것은 무척이나 간편해져서 오늘날 사람들은 자신의 주변 환경을 손쉽게 측정할 수 있게 되었다. 해당 장치는 비용 효율적이고 무엇보다도 아주 빠르게 건물의 3차원 측정을 용이하게 한다. 따라서 평면도, 표면적 및 볼륨을 캡처할 수 있을 뿐만 아니라 부분적으로 매우 상세한 3D 보기를 생성할 수도 있다.
이런 레이저 스캐너의 기본 디자인 개념은 부분적으로 크게 다르다. 광학 CMM 3D 스캐너 MetraSCAN 3D는 전문적인 조건에서 매우 정확한 측정 결과가 필요한 엔지니어의 요구사항에 대응하기 위해 모듈식으로 디자인되었다. 이 시스템을 사용하면 매우 복잡한 데이터를 절대 정밀도로 수집할 수 있다. 반면에 이미지 생성 레이저 스캐너인 Leica BLK360의 디자인은 모든 사람이 사용할 수 있는 정교한 측정기술을 전제로 진행되었다. 복잡한 기능들을 극도로 축소된 형태로 변환한다. 이 스캐너는 하나의 버튼으로 작동하는데, 활성화되면 스캐너가 회전하고 주변을 기록한다. 앱은 실시간으로 스캔 데이터를 필터링하고 등록한다. 이런 데이터를 기반으로 사용자는 가능한 가장 간단한 방법으로 풀 컬러, 고해상도, 3D 파노라마 이미지를 제공받게 된다.

## Leica BLK360

**Imaging Laser Scanner**
이미징 레이저 스캐너

**Manufacturer 제조**
Leica Geosystems AG
헤르브루그 | 스위스
**Design 디자인**
platinumdesign (Matthias Wieser)
슈투트가르트 | 독일
www.leica-geosystems.com

**1**

**1**

## MaxSHOT 3D

**Optical Measuring System**
광학 측정 시스템

**Manufacturer 제조**
Creaform Inc, Lévis
퀘백 | 캐나다
**In-house design 인하우스 디자인**
François Lessard, Nicolas Lebrun
www.creaform3d.com

3

2

## 2
### DeWALT Rotary Laser

**Imaging Laser Scanner**
레이저 스캐너

**Manufacturer 제조**
Stanley Black & Decker
사우딩턴 | 코네티컷 | 미국
**In-house design 인하우스 디자인**
Vincent Cook
**Design 디자인**
Michael Matteo
시카고 | 미국
www.stanleyblackanddecker.com

## 3
### MetraSCAN 3D /
### HandyPROBE / C-Track

**Portable 3D Measurement System**
휴대용 3D 측정 시스템

**Manufacturer 제조**
Creaform Inc. (Ametek Inc.), Lévis
퀘백 | 캐나다
**In-house design 인하우스 디자인**
François Lessard, Nicolas Lebrun
www.creaform3d.com

# Digitalising 디지털화

Whereas photography at its beginning was purely a matter of chemistry and mechanics, today's digital photography is based on highly advanced technology. This makes it possible to capture reality within fractions of a second by means of electronics and light sensitive chips, and convert it into digital images. This technology is found not only in digital cameras, but also in every smartphone and tablet, and is thus available to anyone at any time. In the past ten years, the smartphone in particular has revolutionised how we take photographs. Today, billions of photos are taken every day, an average of about 800 photos are posted on Instagram alone – every second[1]. At the same time, the cameras are merely what stand out periscope-like in an enormous, undulating sea of digital image data and supply new photos. There are already many new devices and services that are able to link photos with additional information such as coordinates, weather information, or facial recognition. Incredible amounts of visual data therefore flow back and forth between devices, hard drives and storage discs, social media and Cloud services, giving rise to a sea of image data.

By far the majority of photos are shot with smartphones. Yet the market for classical digital, compact or system cameras is rebounding. Whoever picks up such a camera wants to take more than just snapshots. This appreciation of photography can be seen in the design of the camera models: many of them are constructed of high quality materials, precisely engineered, and perfectly balanced. This is also true, for instance, in the case of the system camera Olympus Pen, which pays homage to analogue photography in its design, but conceals the most modern optical technology inside. Ease of operation also plays a central role for all cameras. User friendliness or usability is important so as to make it possible to use all the technology contained in the camera intuitively.

[1] Source: www.internetlivestats.com

사진의 시작은 순전히 화학과 기계적인 작동이었으나 오늘날의 디지털 사진은 고도로 발전된 기술들을 기반으로 한다. 이를 통해 전자 장치와 감광성 칩을 사용하여 몇 초 안에 현실을 포착하고 디지털 이미지로 변환할 수 있다. 이 기술은 디지털카메라는 물론 모든 스마트폰과 태블릿에서 볼 수 있어 언제 어디서나 누구나 사용할 수 있다. 지난 10년 동안 스마트폰은 우리가 사진을 찍는 방식에 혁명을 일으켰다. 오늘날 매일 수십억 장의 사진이 찍히고 인스타그램에만 1초에 평균 약 800장의 사진이 게시된다. 동시에 카메라는 거대한 디지털 이미지 데이터의 바다에서 잠망경처럼 눈에 띄고 새로운 사진을 제공하는 것일 뿐이다. 사진을 좌표, 날씨 정보 또는 얼굴 인식과 같은 추가 정보와 연결할 수 있는 새로운 장치와 서비스가 이미 많이 있다. 따라서 엄청난 양의 시각적 데이터가 장치, 하드 드라이브 및 스토리지 디스크, 소셜 미디어 및 클라우드 서비스 사이를 오가며 이미지 데이터의 바다를 형성한다.

지금까지 대부분의 사진은 스마트폰으로 촬영되었다. 그러나 기존의 디지털, 컴팩트 또는 시스템 카메라 시장이 다시 회복되고 있다. 그런 카메라를 집어 든 사람은 스냅샷 그 이상을 찍고 싶어 한다. 사진에 대한 이런 고마움은 카메라 모델의 디자인에 나타나 있다. 많은 모델이 고품질 재료로 제작되고 정밀하게 디자인되었으며 완벽하게 균형이 잡혀 있다. 예를 들어, 시스템 카메라인 Olympus Pen의 경우에도 마찬가지다.

**360-Degree Camera**
360도 카메라

**Manufacturer 제조**
Memora Inc.
팔로 알토 | 캘리포니아 | 미국
Memora APAC Co., Ltd.
타이베이 | 대만
**In-house design 인하우스 디자인**
Ju-Chun Ko, Chiang Fon,
Sunny Chou, Tai-Pei Wang,
Dora Lee, Jackie Chia-Hsun Lee,
Servando Canales, Gaurav Gupta,
Yi-Hao Yeh, Kevin Lin
www.luna.camera

Thanks to rapidly developing virtual reality technologies, ever more manufacturers are bringing 360-degree cameras onto the market. These differ from normal cameras above all in shape. Their, for the most part, round form is derived from the many camera modules that shoot the environment in a 360-degree range. The IR camera Ozo with its eight cameras is also spherical in design and has the futuristic aesthetics that we are familiar with from science fiction films.

디자인은 아날로그 사진에 대한 나름대로의 경의를 표하고 있지만, 내부에는 가장 현대적인 광학 기술이 숨겨져 있다. 조작의 용이성 또한 모든 카메라에서 중심적인 역할을 한다. 카메라에 포함된 모든 기술을 직관적으로 사용할 수 있도록 하기 위해서는 사용자의 편의성이나 사용성이 중요하다.

빠른 속도로 발전하는 가상 현실 기술에 힘입어 360도 카메라를 출시하고 있는 제조업체가 늘어나고 있다. Luna 카메라는 우선 일반 카메라와 모양이 다르다. 공처럼 둥근 형태는 360도로 촬영하는 카메라 모듈에서 가져온 것이다. 여덟 개의 카메라가 장착된 IR 카메라 Ozo도 그런 구형 디자인을 채택한 경우로, 공상과학 영화에서나 볼 수 있는 미래 지향적인 모습을 하고 있다.

# Blackmagic Pocket Cinema Camera

**Compact Digital Cinema Camera**
컴팩트 디지털 시네마 카메라

**Manufacturer 제조**
Blackmagic Design Pty Ltd
호주
**In-house design 인하우스 디자인**
Blackmagic Industrial Design Team
호주
www.blackmagicdesign.com

**1**

**2**

**5**

**1**

Ozo

**Virtual Reality Camera**
가상 현실 카메라

**Manufacturer 제조**
Nokia Technologies
서니베일 | 캘리포니아 | 미국
**In-house design 인하우스 디자인**
Nokia Design
www.nokia.com

**2**

OCLU

**Action Camera**
액션 카메라

**Manufacturer 제조**
OCLU Limited
런던 | 영국
**In-house design 인하우스 디자인**
Hugo Martin
www.oclu.com

**3**

PEN-F

**System Camera**
시스템 카메라

**Manufacturer 제조**
Olympus Europa SE & Co. KG
함부르크 | 독일
**In-house design 인하우스 디자인**
Olympus Corporation Design
Center
(Keji Okada, Takeshi Nohara)
도쿄 | 일본
www.olympus-global.com

**4**

Blackmagic Micro
Camera

**Ultra-Compact Digital Film Camera**
초소형 디지털 필름 카메라

**Manufacturer 제조**
Blackmagic Design Pty Ltd
멜버른 | 호주
**In-house design 인하우스 디자인**
www.blackmagicdesign.com

**4**

**3**

**6**

## 5
### KeyMission 360

**Action Camera**
액션 카메라 (액션캠)

**Manufacturer 제조**
Nikon Corporation
도쿄 | 일본
**In-house design 인하우스 디자인**
Hiroyuki Asano, Chikara Fujita,
Mitsuo Nakajima
www.nikon.com

## 6
### Canon PowerShot G9X

**Compact Camera**
콤팩트 카메라

**Manufacturer 제조**
Canon Inc.
도쿄 | 일본
**In-house design 인하우스 디자인**
www.canon.com

Obse
P
관

# PowerEye

**Quadrocopter**
쿼드콥터

**Manufacturer 제조**
PowerVision Technology Co., Ltd.
베이징 | 중국
**In-house design 인하우스 디자인**
Brian Yuan
www.powervision.me

## Observing 관찰

Drones, originally developed as flying practice targets for military uses, hold enormous potential: They extend people's operating range by being remotely controlled or moving autonomously and hence create distance between the person and the action that he or she has initiated. This distancing can be problematic, but is nonetheless helpful in many cases. As transport drones, for example, the machines can be flown over crisis regions to deliver medication to remote areas without risking lives. And, when equipped with cameras, drones broaden our horizons: They provide us with magical views and insights that could previously only be obtained from conventional aircraft or helicopters. This flight with a bird's-eye view is not only fascinating, but meanwhile available to a broad public as well. Because, even though drones are equipped with ever more complex technologies, they are also becoming ever smaller, lighter, more affordable – and easier to operate – at the same time.

드론은 본래 군사용 비행 연습 표적으로 개발되었지만 그 이상의 엄청난 잠재력이 있다. 드론은 원격으로 조종하거나 자율적으로 이동하는 운용 방식을 통해 인간의 활동 범위를 넓혀주어, 멀리서도 작업을 수행하는 것이 가능하다. 이러한 거리두기는 문제가 될 수 있지만 도움이 되는 경우가 많다. 예를 들어, 위험한 지역 너머 외진 곳에 약품을 전달해야 하는 경우, 전달하는 사람의 생명에 지장 없이 운송용 드론으로 처리할 수 있다. 카메라가 장착된 드론은 사람의 시야를 넓혀준다. 과거의 항공기나 헬리콥터에서만 볼 수 있었던 멋진 경관을 이제는 손쉽게 볼 수 있고 대상을 더 잘 알아볼 수 있게 해준다. 높은 곳에서 아래를 바라보며 비행하는 경험은 매력적이기도 하고, 누구나 조작할 수 있다. 드론에는 점점 더 복잡한 기술이 적용되면서도 동시에 점점 더 작고, 가볍고, 저렴하고, 쉽게 작동할 수 있게 개발되기 때문이다.

훈련을 받은 사람은 물론, 민간 용도로도 누구나 쿼드콥터나 멀티콥터를 이용할 수 있게 된 지 오래다. 이렇게 되기까지 디자이너들이 큰 역할을

For a long time now, it has been possible for anyone to control such quadro- or multicopters for civil applications, not only trained operators. Designers have played a big role in this development: They have designed human-machine interaction and thus contributed to the fact that these flying robots, which move through the air around us and over our heads, are as easy as possible to operate.

The design of the Hover Camera, an autonomous minicopter fitted with a camera for recreational use, which is optimised above all for selfies, is particularly successful in terms of interaction. The Hover Camera can be steered playfully by means of gestures and smartphone and simply caught in the hand, since the rotors of this extremely light quadrocopter are protected inside a sturdy, cage-like housing. Integrated AI technology also employs face and body recognition so that, when in autonomous mode, the drone swirls around users or follows them like an object in standby. The Hover Camera thus offers seemingly natural, emotionalised human-machine interaction that guarantees a great degree of control – and is therefore lots of fun.

했다. 공중을 가르며 머리 위로 날아다니는 이 로봇을 가능한한 쉽게 조작할 수 있도록 인간과 기계의 인터렉션을 디자인했다.

특히 셀카에 최적화된 레크리에이션용 카메라가 장착된 자율 운행 미니 콥터인 Hover Camera의 디자인은 상호작용 측면에서 매우 성공적이다. Hover Camera는 제스처와 스마트폰을 사용하여 재미있게 조종할 수 있으며, 이 초경량 쿼드콥터의 날개는 철창같이 생긴 견고한 덮개 내부에서 작동하기 때문에 손으로 잡을 수도 있다. 통합 AI 기술도 얼굴 및 신체 인식을 사용하여 자율 운행 모드에서는 마치 드론이 대기하고 있는 것처럼 사용자 주위를 빙빙 돌거나 따라오게 할 수 있다. 이런 기능으로 인해 Hover Camera는 고도로 제어할 수 있는 자연스럽고 감성적인 인간-기계 인터렉션을 제공하므로 매우 재미있다.

## Hover Camera

**Drone**
드론

**Manufacturer 제조**
Zero Zero Robotics Inc.
베이징 | 중국
**In-house design 인하우스 디자인**
Mengqiu Wang, Guanqun Zhang
www.gethover.com

**1**

## PowerEgg

**Quadrocopter**
쿼드콥터

**Manufacturer 제조**
PowerVision Robot Inc.
베이징 | 중국
**In-house design** 인하우스 디자인
Feng Fan
www.powervision.me

**2**

## GoPro Karma System

**Drone**
드론

**Manufacturer 제조**
GoPro, Inc., San Mateo
샌머테이오 | 캘리포니아 | 미국
**In-house design** 인하우스 디자인
GoPro Industrial Design Team
www.gopro.com

**3**

## PHANTOM 4 PRO

**Drone**
드론

**Manufacturer 제조**
DJI
선전 | 중국
**In-house design** 인하우스 디자인
www.dji.com

**4**

## Inspire 2

**Drone**
드론

**Manufacturer 제조**
DJI
선전 | 중국
**In-house design** 인하우스 디자인
www.dji.com

3

5

6

## 5
## Airblock

**Drone for Education**
교육용 드론

**Manufacturer 제조**
Makeblock Co., Ltd.
선전 | 중국
**In-house design 인하우스 디자인**
Mingzhe Guo, Jimmy Qin,
Wenhua Wang
www.makeblock.com

## 6
## MAVIC PRO

**Drone**
드론

**Manufacturer 제조**
DJI
선전 | 중국
**In-house design 인하우스 디자인**
www.makeblock.com

Explo

탐
□

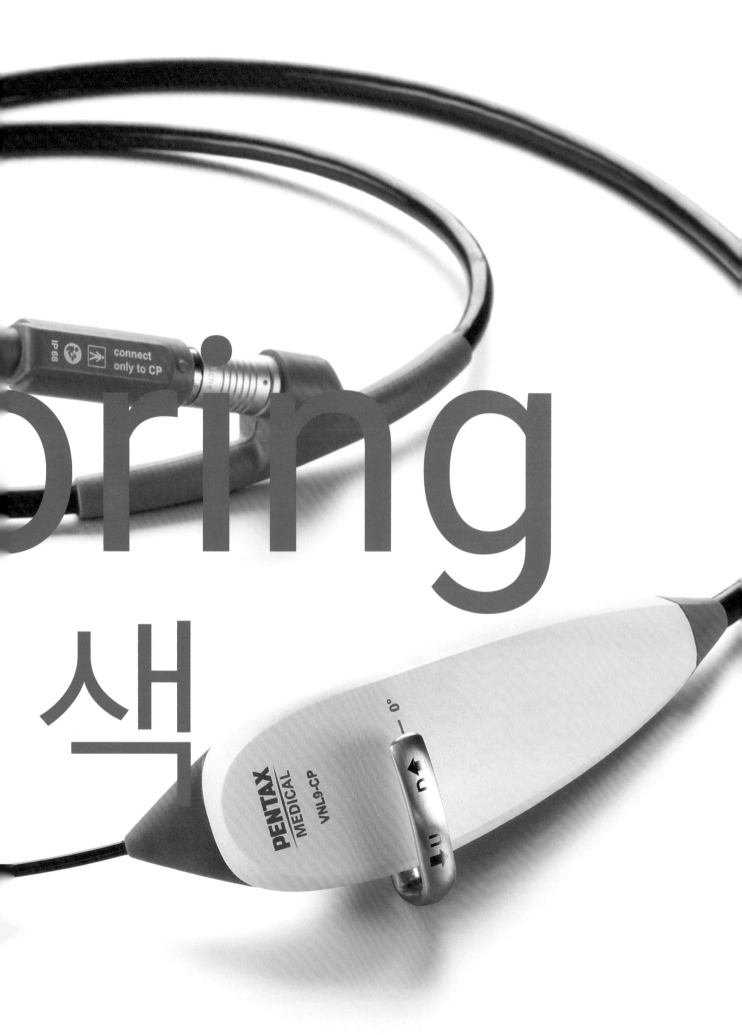

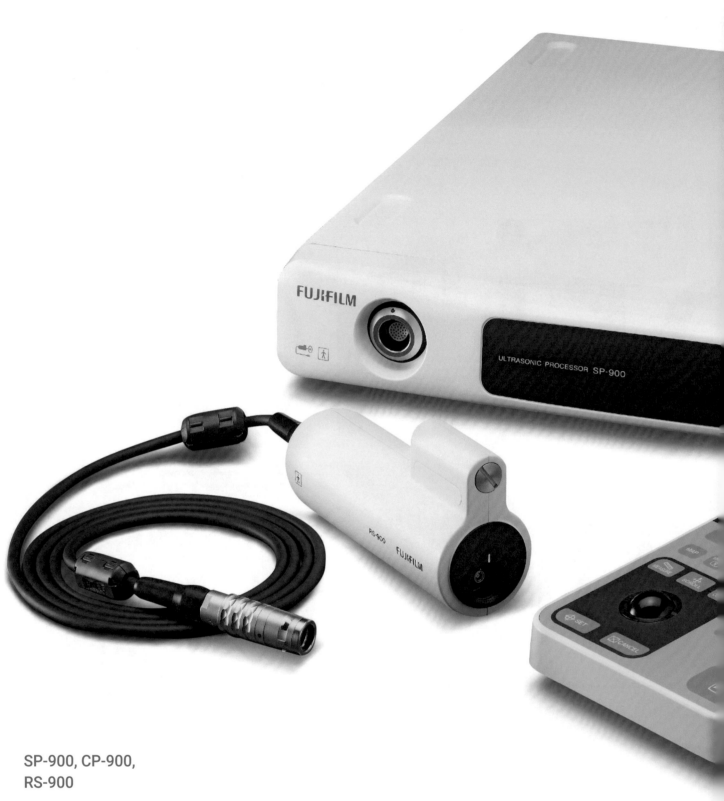

SP-900, CP-900,
RS-900

**Endoscopic Ultrasound
Probe System**
내시경 초음파 시스템

**Manufacturer** 제조
FUJIFILM Corporation
도쿄 | 일본
**In-house design** 인하우스 디자인
Kunihiko Tanaka
www.fujifilm.com

## Exploring 탐색

While digital cameras, action cameras, and 3D cameras document images of the world as we see and know it, in the field of medicine, digital imaging generates images of a world that is usually hidden – the inside of the body. This form of imaging includes, for example, computer tomography, magnetic resonance tomography, X-ray images, and endoscopy. Thanks to this digitalisation in medical technology, storing and further processing these image data have become considerably simpler. On the monitor screen, doctors can zoom into the images in real time, measure areas and abnormalities using corresponding software, and prepare initial findings. Above all the evaluation of images with the support of artificial intelligence will, at the same time, considerably alter and ease the work of doctors in the years to come, since, even today, it already partially facilitates comparing images with huge databases, thus reducing the danger of incorrect medical diagnoses as well as helping identify optimal treatment plans.

디지털카메라, 액션 카메라 및 3D 카메라는 세상을 우리가 보고 알고 있는 그대로 기록하지만, 의학 분야에서의 디지털 이미징은 일반적으로 몸 안의 보이지 않는 부분을 보여준다. 이런 형태의 영상 촬영에는 컴퓨터 단층 촬영, 자기 공명 단층 촬영, 엑스레이, 내시경 검사 등이 포함된다. 이런 의료 기술의 디지털화로 인하여 이런 영상 데이터를 저장하고 추가로 처리하는 일이 훨씬 더 간단 해졌다. 의사는 모니터 화면을 통해 실시간으로 영상을 확대하고 관련 소프트웨어를 사용하여 면적과 이상부위를 측정하고 초기 소견을 작성할 수 있다. 무엇보다도 인공지능을 이용한 영상 평가는 향후 의사가 하는 일에 큰 변화를 가져옴과 동시에 쉬워질 것이다. 지금도 이미 막대한 데이터베이스와 영상을 비교하는 일이 부분적으로 간편해졌으며, 잘못된 의료진단의 위험을 감소하는 것은 물론 최

The basis of all this is imaging itself. And this too has changed thanks to the digitalisation of medical imaging. In modern endoscopy today, video endoscopy that makes use of digital technologies for imaging and transmits the images to the monitor screen via glass fibre image conductors is therefore used. These endoscopes are sensitive precision instruments that have to be designed in a correspondingly sensitive way so as to ensure safe handling. In the case of the Video Coloscope EC-760ZP-V/L for examining the bowel, for example, improved power transmission is ensured by the fact that the device has a head that can be moved freely. Colour coding on the handle helps avoid possible operating errors. The Vivideo endoscope for diagnostic examinations of the ear, nose, and throat area, on the other hand, offers above all greater patient comfort thanks to the combination of ultra-flexible material and the small diameter of the insertion tube. The handle, which calls to mind a humming bird, is coated with silicon and allows the examiner to hold it naturally.

1

적화된 치료 계획을 세우는 데 도움이 되고 있기 때문이다.

이 모든 것의 기초가 되는 것이 이미징이다. 그리고 이 역시 의료영상의 디지털화로 인하여 바뀌어서 오늘날 현대적인 내시경검사에서는 디지털 기술을 이용하여 영상을 촬영하고 유리 섬유 영상 전도체를 통해 모니터 화면으로 영상을 전송하는 영상 내시경이 사용된다. 이런 내시경은 민감한 정밀기기로써 안전한 사용을 보장하기 위해 그에 맞게 민감하게 설계되어야 한다. 예를 들어, 장 검사용 Video Coloscope EC-760ZP-V/L의 경우, 자유롭게 움직일 수 있는 머리 부분이 있어 향상된 동력의 전달을 가능하게 했다. 색상으로 손잡이를 구분한 것은 사용상 오류를 미연에 방지하는 데 도움이 된다. 이비인후과 진단 검사를 위한 Vivideo 내시경은 매우 유연한 소재로 제작한 직경이 작은 삽입 튜브를 사용해 환자가 더 편안하게 검사를 받게 한다. 벌새를 연상시키는 손잡이는 실리콘으로 코팅해 검사자가 자연스럽게 손에 쥘 수 있게 했다.

## 1

### Q-tube Wi-Fi Teeth Scope Pro | Q-tube Wi-Fi Otoscope Pro

**Teeth Scope and ENT Otoscope**
치아 스코프, 이비인후과 검이경

**Manufacturer 제조**
Quanta Computer Inc.
타오위안 | 대만
**In-house design 인하우스 디자인**
Po-Hsian Tseng, Chu-Fu Wang,
Chang-Ta Miao, Guo-Chyuan Chen
www.quanta.com.tw

## 2

### EC-760ZP-V/L

**Video Endoscope**
비디오 내시경

**Manufacturer 제조**
FUJIFILM Corporation
도쿄 | 일본
**In-house design 인하우스 디자인**
Kunihiko Tanaka
www.fujifilm.com

2

3

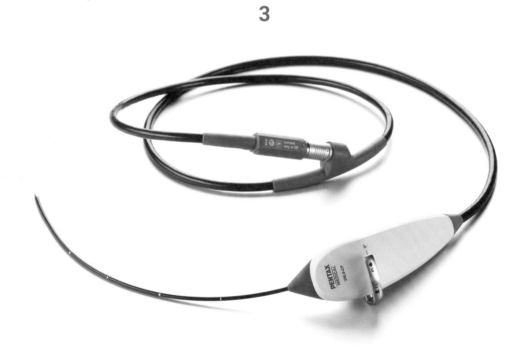

## 3
## Vivideo

**ENT Video Endoscope**
이비인후과 비디오 내시경

**Manufacturer 제조**
Pentax Medical
프리트베르크 | 독일
**In-house design 인하우스 디자인**
Christoph Rilli
www.pentaxmedical.com

Proce

ssing

# Processing 처리

In the past, all the data that are today collected in digital form – as described in the previous chapters – were processed, evaluated, and translated into actions only by people. In the age of digitalisation, however, people no longer only rely on their own intelligence. Instead, data are fed into Big Data systems, cross-checked with other data, put in relation to each other, dissected by algorithms, sorted anew, and evaluated. Today, these networked systems are so complex that they far exceed the comprehension of a single person and, paradoxically, are nonetheless able to provide diagnoses or recommendations for action that are much more individual. The sheer quantity of (comparative) data alone makes human beings calculable in the truest sense of the word. At the same time, the data make it possible for people to perform highly complex calculations. People are therefore simultaneously subject and object in a continuous data transfer. The intermediaries between human beings and Big Data are, in turn, mainframe computers and servers, personal computers, tablets and smartphones, wearables, storage devices, and intelligent devices and machines. Such devices are black boxes within which complex processes take place. These black boxes got their start in the development of cameras. All at once, reality was no longer reproduced by hand in the form of a drawing, but depicted true-to-reality by complex physical and chemical processes inside the camera, without direct intervention by human beings. Everything that happened between pressing the shutter release and the finished image escaped the observation and knowledge of by far the majority of people, thus lending the device a mysterious aura. Concealed in today's black boxes are calculations, algorithms, data clouds, access to the internet, virtual reality, and neuronal networks –
in short, Big Data. All this can potentially be accessed and used via these black boxes.

전에는 이전 장에서 설명한 것처럼 현재 디지털 형식으로 수집되는 모든 데이터를 처리, 평가하고 그에 따른 행동을 취하는 것은 사람만이 할 수 있었다. 그러나, 디지털화 시대의 사람들은 더 이상 자신의 지능에만 의존하지 않는다. 데이터는 빅 데이터 시스템에 입력되고, 다른 데이터와 교차 확인되고, 데이터 사이의 관계를 정하고, 알고리즘으로 나뉘고, 다시 정렬되고, 평가된다. 오늘날 이런 네트워크 시스템은 너무 복잡해서 한 사람이 이해할 수 있는 수준을 넘어서는데, 그럼에도 불구하고 역설적이게도 훨씬 더 개인적인 조치에 대한 진단이나 권고사항을 제공할 수 있다. 순전히 (비교) 데이터의 양만으로도 인간은 말 그대로 예측이 가능하다. 동시에 사람들은 데이터를 이용해 매우 복잡한 계산을 할 수 있다. 그러므로 사람은 지속적인 데이터 전송에 있어 주체이자 객체가 된다. 인간과 빅 데이터 사이의 매개체는 결국 중앙 컴퓨터와 서버, 개인용 컴퓨터, 태블릿과 스마트폰, 웨어러블, 저장장치, 지능형 장치와 기계가 된다. 이런 장치들은 복잡한 프로세스가 발생하는 블랙박스 (복잡한 기계장치)라고 한다. 이런 블랙박스는 카메라를 개발하는 과정에서 시작되었다. 어느 순간부터 더 이상 손으로 그리는 대신 카메라 내부의 복잡한 물리적, 화학적 과정을 통해 인간의 직접적인 개입 없이 현실을 사실대로 묘사하기 시작했다. 셔터를 누를 때부터 사진이 완성되기까지 일어나는 모든 일은 일반 대중의 관찰과 지식으로는 이해하기 어려운 것이 되었고, 장치에 신비감이 생기게 되었다. 오늘날의 블랙박스에는 계

## ROG Maximus IX APEX

**Motherboard**
마더보드

**Manufacturer 제조**
ASUSTeK Computer Inc.
타이베이 | 대만
**In-house design 인하우스 디자인**
www.asus.com

This complexity inside devices is juxtaposed today with the most purist possible design of the housing, as, for example, in the case the HP Elite Slice Desktop Computer. Things generally first become really exciting when the design of machines breaks with this purism. For instance in the field of gaming, where the performance of the individual components is of utmost importance and users therefore like to assemble the optimal computer for their requirements themselves, while both housing as well as elements such as motherboard or graphic card express power and aggression as a result of their distinctive design. One example of this is the motherboard ROG Maximus IX APEX.

산, 알고리즘, 데이터 클라우드, 인터넷 액세스, 가상 현실, 뉴런 네트워크, 즉 빅 데이터가 숨겨져 있다. 이 모든 것은 블랙박스를 통해 접근하고 사용할 수 있다.

이런 장치 내부의 복잡성은 오늘날 HP Elite Slice 데스크탑 컴퓨터의 경우와 같이 가장 순수한 케이스 디자인과 비교된다. 기계의 디자인은 이런 순수주의에서 벗어나게 되면 모든 것이 흥미로워진다. 예를 들어, 게임 분야에서는 컴퓨터 내 개별 부품의 성능이 무엇보다도 중요하기 때문에 사용자 스스로 자신의 요구사항에 맞게 최적의 컴퓨터를 조립하는 것을 선호하는 데, 케이스, 메인보드나 그래픽 카드와 같은 부품 모두 독특한 디자인을 통해 강력함과 적극성을 표현한다. 이와 관련된 사례로는 메인보드 ROG Maximus IX APEX가 있다.

## HP Elite Slice

**Computer**
컴퓨터

**Manufacturer 제조**
HP Inc.
팔로 알토 | 캘리포니아 | 미국
**In-house design 인하우스 디자인**
HP Design
**Design 디자인**
Native Design
런던 | 영국
www.hp.com

1

2

5

**1**

## Mac Pro

맥 프로

**Manufacturer 제조**
Apple
쿠퍼티노 | 캘리포니아 | 미국
**In-house design 인하우스 디자인**
www.apple.com

**2**

## Lenovo Smart Storage

**Storage System**
스토리지 시스템

**Manufacturer 제조**
Lenovo
모리스빌 | 노스캐롤라이나 | 미국
**In-house design 인하우스 디자인**
www.lenovo.com

**3**

## Raiden Series

**Gaming Memory Modules**
게임용 메모리 모듈

**Manufacturer 제조**
AVEXIR Technologies Corp.
주베이 | 대만
**Design 디자인**
Shenzhen Chocolight Technology
Co., Ltd.
선전 | 중국
www.avexir.com

**4**

## Engine 27

**CPU Cooler**
CPU 쿨러

**Manufacturer 제조**
Thermaltake Technology Co., Ltd.
타이베이 | 대만
**In-house design 인하우스 디자인**
www.thermaltake.com

**3**

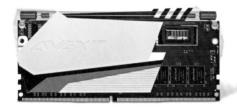

**4**

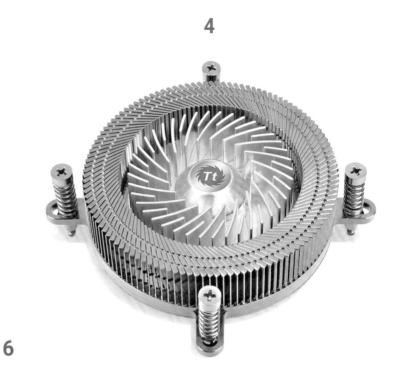

**6**

**5**
## MasterLiquid Maker 92

**CPU Cooler**
CPU 쿨러

**Manufacturer 제조**
Cooler Master Technology Inc.
신베이 | 대만
**In-house design 인하우스 디자인**
www.coolermaster.com

**6**
## ROG Maximus IX
## Extreme

**Motherboard**
마더보드

**Manufacturer 제조**
ASUSTeK Computer Inc.
타이베이 | 대만
**In-house design 인하우스 디자인**
www.asus.com

**1**

**2**

**4**

**1**

## Power Mac G5

**Desktop computer**
데스크톱

**Manufacturer 제조**
Apple, Inc.
쿠퍼티노 | 캘리포니아 | 미국
**In-house design 인하우스 디자인**
Apple Industrial Design Team
www.apple.com

**2**

## HP Envy Phoenix

**Computer**
컴퓨터

**Manufacturer 제조**
HP Inc.
팔로 알토 | 캘리포니아 | 미국
**In-house design 인하우스 디자인**
HP Industrial Design Team
www.hp.com

**3**

## HP Pavilion Mini

**Mini Computer**
미니 컴퓨터

**Manufacturer 제조**
HP Inc.
팔로 알토 | 캘리포니아 | 미국
**In-house design 인하우스 디자인**
HP Industrial Design Team
www.hp.com

**4**

## LaCie Porsche Design Mobile Drive

**External Hard Drive**
외장 하드 드라이브

**Manufacturer 제조**
Seagate Technology
쿠퍼티노 | 캘리포니아 | 미국
**Design 디자인**
Studio F. A. Porsche
Nadine Cornehl
Zell am See | Austria
www.seagate.com

**3**

**5**

**6**

## 5
### Walkman NW-WM1Z

**Portable Audio Player**
휴대용 오디오 플레이어

**Manufacturer 제조**
Sony Video & Sound Products Inc.
도쿄 | 일본
**In-house design 인하우스 디자인**
Sony Corporation (Soichi Tanaka)
도쿄 | 일본
www.sony.net

## 6
### Chromé

**Hard-Drive System**
하드 드라이브 시스템

**Manufacturer 제조**
Seagate Technology
쿠퍼티노 | 캘리포니아 | 미국
**Design 디자인**
Neil Poulton Industrial &
Product Design
파리 | 프랑스
www.seagate.com

# Conn

연

ecting

결

5G 2G

Ø 1 2 3

## Connecting 연결

Digitalisation has entered many areas of our lives – including education and culture, media and politics, business and industry as well as how we communicate and our own four walls. Digitalisation first reveals its full potential through networking. Only then are people, objects, and processes connected with one another and able to exchange data – the basis for Big Data. How self-evident such networking has meanwhile become only becomes truly apparent when it is not available. When a server connection breaks down, we are catapulted back to the pre-internet era in a fraction of a second.

In the exhibition "Homo ex Data", Wi-Fi routers, which ensure connection to the internet, are representative of connectivity. They have become indispensible, but also taken for granted. They are pragmatic devices; the function is of primary importance, both in perception and design. They are supposed to fit inconspicuously into their surroundings and provide their service reliably. As symbols of connectivity, the antennae are the most prominent characteristic of routers.

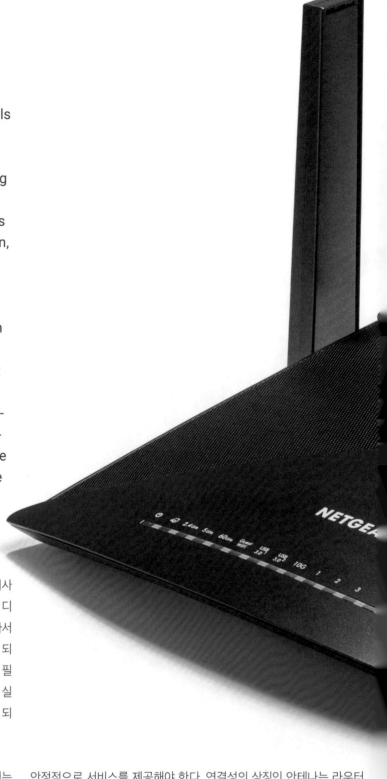

디지털화는 교육과 문화, 미디어와 정치, 비즈니스와 산업은 물론 의사소통 방식과 사적인 공간을 포함하여 삶의 많은 영역에 침투해 있다. 디지털화는 먼저 네트워킹을 통해 완전한 잠재력을 드러낸다. 그러고 나서야 사람, 사물, 프로세스가 서로 연결되어 데이터를 교환할 수 있게 되며 이것이 빅데이터의 기반이 된다. 네트워킹의 영향은 따로 증명할 필요가 없지만, 네트워킹을 할 수 없을 때가 되어야 비로소 그 영향을 실감하게 된다. 서버 연결이 끊어지면 1초도 안되어 인터넷 이전 시대로 되돌아가게 된다.

'Homo ex Data' 전시회에서 인터넷 연결을 담당하는 Wi-Fi 라우터는 연결성을 나타낸다. 라우터는 필수적인 것이 되었지만 당연히 있는 것으로 생각하기도 한다. 라우터는 실용적인 장치이며 그 기능은 지각적으로나 설계적인 면에서도 아주 중요하다. 주변에 눈에 띄지 않게 설치되어 안정적으로 서비스를 제공해야 한다. 연결성의 상징인 안테나는 라우터의 가장 눈에 띄는 특징이다.

### 360 Smart Router P2R

**Smart Router**
스마트 라우터

**Manufacturer 제조**
Shenzhen Fenglian Technology
Co., Ltd.
선전 | 중국
**In-house design 인하우스 디자인**
Shengjiong Zhang,
Xueyong Zhang
www.ifenglian.com
Page 90 | 91

### Nighthawk X10

**Router**
라우터

**Manufacturer 제조**
NETGEAR, Inc.
새너제이 | 캘리포니아 | 미국
**In-house design 인하우스 디자인**
www.netgear.com

Cont

# Swisscom TV 2.0

**Remote Control**
리모컨

**Manufacturer 제조**
Swisscom AG
취리히 | 스위스
**Design 디자인**
ruwido austria gmbh
노이마르크트 | 오스트리아
www.swisscom.ch

# Controlling 제어

For human beings to be able to interact with black boxes and gain access to the data that are stored or processed in them, a human-machine interface is necessary. It is first by means of such an interface that human beings can become part of the Big Data system, enter data, run programs, and retrieve the processed data. The way in which people interact with machines also characterises their relationship with technology. The simpler and more efficient the communication is, the more logical it is for us to use a computer. Over 200 years ago, the first Jacquard loom was already automated by means of punch cards – a type of human-machine communication that also continued to be used in the first computers – although the punch card quickly took over the role of a storage medium. Working with punch cards was laborious and interaction with the machines was restricted to only a few people. The possibility then developed to communicate with the computer by means of a keyboard and entering commands in the command line. After the market launch of Xerox and Apple's graphic interfaces of the 1970s and 1980s, which were based on the computer program Sketchpad developed by Ivan Sutherland, communication with the device became easier and more intuitive. Since then, by using a mouse, trackball, keyboard, graphic tablet, game controller, stylus, or monitor it has become extremely simple to operate a computer, even for people who only have a layperson's understanding of what occurs in this black box. These devices, similar to a television remote control, are, in principle, an extension of the hand with which we are able to reach and manipulate virtual objects on the screen effortlessly, without actually touching them. These manual entry devices are still in use today, supplemented by newer interfaces such as touchscreens or voice control.

인간이 블랙박스와 상호작용하고 블랙박스에 저장되거나 처리되는 데이터에 접근하려면 인간과 기계간의 인터페이스가 필요하다. 이런 인터페이스를 통해 인간이 빅데이터 시스템의 일부가 되어 데이터를 입력하고 프로그램을 실행하고 처리된 데이터를 회수하게 된 것은 전에는 하지 못했던 일이다. 사람들이 기계와 상호작용하는 방식도 사람-기술 간 관계의 특징을 정의한다. 의사소통은 간단하고 효율적일수록 컴퓨터를 사용하는 것이 더 논리적인데, 200년 전, 최초의 자카르 직기는 이미 펀치 카드를 통해 자동화되었지만, 이런 방식으로 노동자가 일하는 것은 어렵고 소수의 사용자만이 할 수 있었다. 초기 컴퓨터의 기능은 일종의 인간-기계 사이의 통신매체로 볼 수 있는데, 점차 키보드를 통해 손쉽게 컴퓨터와 통신하고 명령줄에 명령을 입력할 수 있게 되었다. 1970년대와 80년대 Xerox와 Apple이 Ivan Sutherland가 개발한 컴퓨터 프로그램 Sketchpad를 기반으로 하는 그래픽 인터페이스를 시장에 출시한 후 장치와 더 쉽고 직관적으로 통신할 수 있게 되었다. 이후로 블랙박스 안에서 무엇이 어떻게 돌아가는 지에 대해 잘 모르는 문외한도 마우스, 트랙볼, 키보드, 그래픽 태블릿, 게임 컨트롤러, 스타일러스, 모니터를 사용하여 매우 간단하게 컴퓨터를 조작하게 되었다. 텔레비전 리모콘과 비슷한 이런 장치는 원칙적으로 연장된 팔과 같으며, 실제로 만지지 않고서도 화면 상의 가상 물체를 만지고 조작할 수 있게 해준다. 이런 수동 입력장치는 터치스크린이나 음성 제어와 같은 새로운 인터페이스로 보완되어 오늘날에도 여전히 사용되고 있다.

1

When designing such control devices, the most important thing is ergonomics. They are tools that, in the truest sense of the word, need to sit comfortably in the hand. This becomes particularly clear in the Xbox Elite Wireless Controller for gaming consoles: its design combines individual adaptability to hand size with ergonomically engineered surfaces and details. Gamers can choose between various metal sticks, direction pads and paddles, according to their wishes. Rubberised grip surfaces give the controller additional traction and stability, so that the device can be held comfortably in the hand, even in case of longer gaming sessions.

이런 제어장치를 디자인할 때 가장 중요한 것은 인체공학이다. 이 말은, 손으로 잡기 편해야 한다는 것이다. 게임 콘솔용 Xbox Elite 무선 컨트롤러를 보면 잘 알 수 있는데, 사람마다 다른 손 크기에 대응하고 표면과 디테일을 인체공학적으로 디자인하였다. 게이머는 다양한 스틱, 방향 패드, 패들 옵션 중에서 원하는 대로 선택할 수 있다. 고무로 코팅한 손잡이 표면은 컨트롤러에 마찰력과 안정성을 높여 오랜 시간 게임을 해도 불편하지 않도록 했다.

1

Nvidia Shield
Android TV

**TV Box Set**
TV 박스 세트

**Manufacturer 제조**
Nvidia Corporation
산타클라라 | 캘리포니아 | 미국
**In-house design 인하우스 디자인**
www.nvidia.com

2

## 2
### Xbox Elite Wireless Controller

**Wireless Controller**
무선 컨트롤러

**Manufacturer 제조**
Microsoft Corporation
레드먼드 | 워싱턴 | 미국
**In-house design 인하우스 디자인**
www.microsoft.com/surface

# 1
## RD TCL RC800 Series

**Remote Controls**
리모콘

**Manufacturer 제조**
TCL Multimedia Technology
Holdings Limited
선전 | 중국
**In-house design 인하우스 디자인**
Innovation Center
www.tcl.com

# 2
## Play Collection

**Wireless Mouse**
무선 마우스

**Manufacturer 제조**
Logitech
뉴어크 | 캘리포니아 | 미국
**In-house design 인하우스 디자인**
Logitech Design
www.logitech.com

# 3
## Bamboo Duo

**Stylus for Touchscreens and Paper**
터치 스크린 및 종이용 스타일러스

**Manufacturer 제조**
Wacom Co. Ltd.
도쿄 | 일본
**Design 디자인**
Human Spark (Scott Lehman)
로즈웰 | 미국
www.wacom.com

# 4
## HP Premium
## Keyboard and Mouse

키보드, 마우스

**Manufacturer 제조**
HP Inc.
팔로 알토 | 캘리포니아 | 미국
**In-house design 인하우스 디자인**
HP Design
**Design 디자인**
Native Design
런던 | 영국
www.hp.com

**3**

**5**

# 5
## Dell Latitude 2-in-1

**Two-in-One Notebook**
투인원 노트북

**Manufacturer 제조**
Dell Inc.
라운드 록 | 텍사스 | 미국
**In-house design 인하우스 디자인**
Experience Design Group
www.dell.com

Touch

Mo 1 .03.2

# Touching 터치

While input devices are needed to control conventional computer monitors, interacting with devices equipped with touchscreens consists simply of gestures such as typing, dragging, or swiping with the fingers. Human-machine communication has thus become much more direct and intuitive, which explains the swift triumph of this technology. The touchscreen as an interface has made external input devices such as a keyboard or mouse at least dispensable. It is also increasingly replacing keys, switches, buttons, regulators, etc., since it serves as an interface for networking in the home, for household appliances, industrial machines, and dispensing automats, in motor vehicles, or at points of sale. Critics, however, find fault with it, claiming that interacting via touchscreen does not correspond to natural, human communication behaviour. Accordingly, the future of human-machine interaction will involve completely doing away with physical control elements. Conversational user interfaces are already tending in this direction, for instance interaction via chatbots, which communicate between humans and machines as personal language assistants. Interaction via brain-computer interface, whereby devices can be controlled directly by thought thanks to implanted electrodes, will be even more direct. Should this kind of communication become successful at some point, and even become bidirectional, meaning that the computer can feed its data directly into the human brain, this could possibly be the point at which human beings and artificial intelligence merge. Until these technologies have truly matured and represent a genuine alternative to touch interfaces, it is up to designers to optimise products with touchscreens as much as possible for each of their fields of application.

One such device that is designed with focus on its

기존의 컴퓨터 모니터를 제어하려면 입력 장치가 필요했지만 터치스크린이 장착된 장치는 그저 손가락으로 타이핑하거나, 드래그 또는 스와이프 같은 제스처를 하기만 하면 된다. 따라서 인간-기계 통신이 훨씬 더 직접적이고 직관적이 되었고, 이 기술이 빠르게 성공한 이유를 설명해 준다. 터치스크린이 인터페이스로 사용되면서 키보드나 마우스와 같은 외부 입력장치가 없어도 되는 장치가 되었다. 또한 터치스크린은 홈 네트워킹, 가전제품, 산업기계, 자판기, 차량 또는 매장에서 인터페이스 역할을 하기 때문에 열쇠, 스위치, 버튼, 조절기 등을 점점 더 많이 대체하고 있다. 그러나 비평가들은 터치스크린을 통한 상호작용이 인간의 자연스러운 의사소통 행위에 해당하지 않는다며 비난하고 있다. 인간-기계 상호작용의 미래는 물리적 제어 요소를 완전히 제거하는 것과 관련이 있다. 대화형 사용자 인터페이스는 이미 이런 방향으로 발전하고 있는데, 사람과 기계 사이에서 의사소통을 담당하는 챗봇을 이용한 상호작용을 예로 들 수 있다. 이식된 전극을 통해 생각으로 직접 장치를 제어할 수 있는 뇌-컴퓨터 인터페이스를 통한 상호작용은 훨씬 더 직접적인 방법이다. 이런 유형의 의사소통이 언젠가 실현되고 컴퓨터가 데이터를 인간의 두뇌에 직접 전달할 수 있는 양방향 소통이 가능해지면 인간과 인공지능이 융합되는 순간이 올 수도 있다. 이런 기술이 크게 향상하고 터치 인터페이스에 대한 진정한 대안이 될 때까지는, 각 응용 분야에서 터치스크린이 있는 제품을 가능한 한 최적화하는 일은 디자이너에게 달려 있다.

## Smart Commander

**Decoding Network Keyboard**
네트워크 디코딩 키보드

**Manufacturer 제조**
Zhejiang Dahua Technology Co., Ltd.
항저우 | 중국
**In-house design 인하우스 디자인**
Yahui Liu, Li Chen
www.dahuasecurity.com

**1**

particular application is the Signature Pad STU-540. This compact pad has a high-resolution, 5-inch colour LCD screen. The tempered, anti-glare glass surface provides a smooth writing sensation and the stylus registers 1,024 pressure ranges. As a result, the pad offers great security when gathering handwritten signatures electronically. The touchscreen of the Busch-Welcome device is used in a completely different context – door communication in private homes. The door communication system, with its 5-inch touch display in vertical format does not take up much space on the wall. Despite its compact form, the system has extensive functions and its operation is intuitive.

특정 용도에 맞춰 디자인된 장치 중 하나는 Signature Pad STU-540이다. 이 작은 패드에는 5인치 고해상도 컬러 LCD 스크린이 있다. 눈부심 방지 처리된 강화유리 표면은 부드러운 필기감을 제공하며 스타일러스는 1,024 단계 압력 감도를 지원하므로, 전자서명을 입력 받을 때 뛰어난 안정감을 준다. Busch-Welcome의 터치스크린은 개인 주택에서의 도어 커뮤니케이션 (대문 앞 의사소통)이라는 완전히 다른 상황에서 사용된다. 5인치 수직형 터치 디스플레이가 있는 도어 커뮤니케이션 시스템은 벽에 많은 공간을 차지하지 않는다. 시스템은 작지만 다양한 기능을 갖추고 있으며 직관적으로 조작할 수 있다.

**2**

**3**

**1**

## JooN3

**Smartwatch**
스마트워치

**Manufacturer 제조**
Infomark
성남 | 대한민국
**Design 디자인**
Purplerain product
안양 | 대한민국
www.infomark.co.kr

**2**

## Busch-Welcome

**Door Communication System**
도어 커뮤니케이션 시스템

**Manufacturer 제조**
Busch-Jaeger Elektro GmbH
member of the ABB-Gruppe
뤼덴샤이트 | 독일
**In-house design 인하우스 디자인**
www.busch-jaeger.de

**3**

## STU-540

**Signature Pad**
서명 패드

**Manufacturer 제조**
Wacom Co. Ltd.
도쿄 | 일본
**In-house design 인하우스 디자인**
Mitchell Giles
www.wacom.com

# Communicating 통신

In smartphones, data input, processing, and output are merged into a single, compact unit, whereby its most distinctive and, as we perceive it, all-dominating design characteristic is the touchscreen, i.e. its interface. The first true smartphone, the iPhone, was introduced in 2007. In the Red Dot Design Yearbook, the new product was described as follows: "The iPhone however is treading new paths, because it combines three products – a mobile phone, a widescreen iPod and an internet-ready handheld computer all in one handy device – thus creating an entirely new type of product. [...] One of the most impressive innovations of the iPhone, however, lies in the approach of how its 3.5" multi-touch screen is operated. It allows users to control the iPhone with just a tap, flick or pinch of their fingers and one can even 'leaf through' the menus: one simply has to swipe the finger gently across the screen in order to open a list for example." At the end of the text, the jury comes to the conclusion:

"[...] The iPhone thus not only lends multimedia communication a new ease of use, it gives a new attitude to life." The iPhone has indeed revolutionised the relationship between human being and computer. Today, life without such a smartphone – regardless of manufacturer – is barely conceivable. It is, so to speak, an interface that has become a product – thanks to gesture control, software, and user-friendly apps, communication between human and computer all at once became simple and intuitive. The incorporation of voice control with the aid of a language assistant such as Siri (first available in the iPhone 4s, 2011) has further simplified interaction with the smartphone and made it more natural.

스마트폰은 데이터 입력, 처리, 출력을 하나의 작은 장치로 결합하는데, 여기서 가장 독특하고 우리가 인식할 때 가장 두드러진 설계적 특징은 스마트폰의 인터페이스인 터치 스크린이다. 진정한 세계 최초 스마트폰인 iPhone은 2007년에 출시되었다. 당시 레드닷 디자인 연감에서는 이 신제품에 대해 이렇게 설명하였다. "그러나 iPhone은 새로운 미래를 열고 있다. 휴대전화, 와이드 스크린 아이팟, 인터넷이 가능한 휴대용 컴퓨터, 이렇게 세 가지 제품을 하나의 편리한 장치에 결합하여 완전히 새로운 유형의 제품을 만들었기 때문이다. [...] 그러나 iPhone이 주도한 가장 인상적인 혁신은 3.5인치 멀티 터치스크린의 작동방식에 있다. 이를 통해 사용자는 탭, 플릭, 핀치만으로 iPhone을 제어할 수 있으며 메뉴 '훑어보기'가 가능하기 때문이다. 예를 들어 목록을 열려면 손가락으로 화면을 쓸어내리기만 하면 된다." 글의 마지막 부분에서 배심원단은 이런 결론을 내린다.

"[...] 그러므로 iPhone은 멀티미디어 통신의 사용 편의성을 높이는 것은 물론 새로운 생활방식을 제시한다." iPhone은 실제로 인간-컴퓨터 관계에 대변혁을 가져왔다. 오늘날 제조업체에 상관없이 스마트폰이 없는 생활은 거의 상상할 수 없다. 말하자면 인터페이스가 제품이 된 것인데, 제스처 제어, 소프트웨어, 사용자 친화적인 앱 덕분에 인간과 컴퓨터 간의 소통이 갑자기 간편하고 직관적으로 변하게 되었다. 음성 제어기능과 시리(2011년 iPhone 4s에서 처음 도입) 같은 언어 도우미 기능의 통합으로 스마트폰과의 상호작용이 더욱 단순하고 자연스러워졌다.

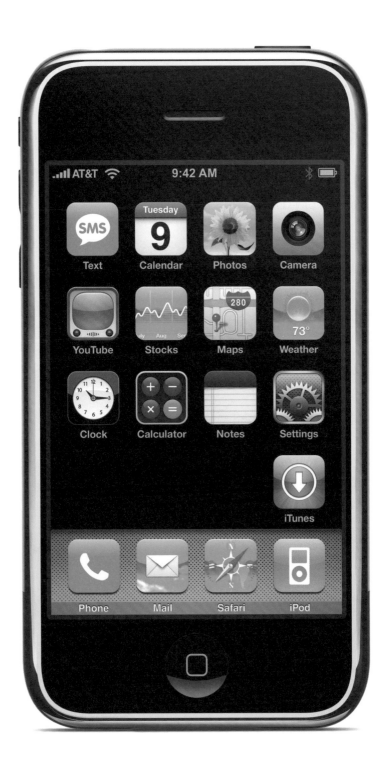

# iPhone (2007)

**Smartphone**
스마트폰

**Manufacturer 제조**
Apple Inc.
쿠퍼티노 | 캘리포니아 | 미국
**In-house design 인하우스 디자인**
Apple Industrial Design Team
www.apple.com

With the first iPhone, Apple succeeded in setting standards not only from a technological, but also from a design viewpoint. Naturally, the iPhone was, in the broadest sense, a further development of existing mobile phones and early smartphones. However, due to the innovative technology, a new archetype needed to be found for this new product class. That the designers succeeded impressively is proven by the fact that, even ten years later, in principle all smartphones adhere to the archetypical design of the iPhone: they must be very flat, possess a touchscreen as big and as frameless as possible, and, preferably, a seamless housing. Physical buttons have been reduced to a minimum and discreetly integrated in the frame. Furthermore, the iPhone 7 Plus visually differs from the original model only marginally. The reason for this is that the classic product design of the smartphone plays, at most, a subordinate role, because what really matters in the case of a smartphone anyway takes place inside the product and on the touchscreen.

**1**

첫 번째 iPhone 출시로 Apple은 기술 표준 및 디자인 표준이 되었다. 물론 iPhone은 포괄적으로 볼 때 기존의 휴대전화와 초기 스마트폰에서 조금 더 발전한 것이다. 그런데 혁신적인 기술로 인해 이 새 제품군에 대한 새 원형을 마련해야 했다. 가능한 한 평평해야 하고, 크고 프레임이 없는 터치스크린이 있어야 하며, 하우징에 이음매가 없으면 더 좋다는 iPhone의 디자인 원형을 십년이 지나도 원칙적으로 모든 스마트폰이 고수하는 것을 보면 디자인이 매우 성공적이었다는 것을 증명해준다. 물리적 버튼은 최소한으로 줄이고 프레임에 눈에 띄지 않게 통합되었다. 게다가 iPhone 7 Plus도 시각적으로는 원 모델과 근소하게 다르다. 스마트폰의 고전적 디자인은 아무리 해봐야 부수적인 역할을 하며, 어쨌든 스마트폰에서 제일 중요한 것은 제품 내부 구성품과 터치스크린이기 때문이다.

**1**

**HTC Desire 530**

Smartphone
스마트폰

**Manufacturer 제조**
HTC
신베이 | 대만
**In-house design 인하우스 디자인**
Daniel Hundt
www.htc.com

**2**

**3**

**4**

**5**

## 2

### arrows NX F-01J

**Smartphone**
스마트폰

**Manufacturer 제조**
Fujitsu Connected Technologies Limited
가와사키 | 일본
**In-house design 인하우스 디자인**
Fujitsu Design Limited
(Kohei Okamoto, Kentaro Yoshihashi)
**Design 디자인**
Imagica Digitalscape Co., Ltd.
(Mai Ichimura, Kensuke Iizuka)
도쿄 | 일본
www.fujitsu.com

## 3

### iPhone 7 Plus

**Smartphone**
스마트폰

**Manufacturer 제조**
Apple Inc.
쿠퍼티노 | 캘리포니아 | 미국
**In-house design 인하우스 디자인**
Apple Industrial Design Team
www.apple.com

## 4

### EF500 Series

**Smartphone**
스마트폰

**Manufacturer 제조**
Bluebird Inc.
서울 | 대한민국
**In-house design 인하우스 디자인**
Bluebird Design Group
(Junk Sik Park)
www.mypidion.com

## 5

### Xperia XA

**Smartphone**
스마트폰

**Manufacturer 제조**
Sony Mobile Communications Inc.
도쿄 | 일본
**Design 디자인**
Sony Corporation
(Keita Hibi, Alexander Sjöstedt)
도쿄 | 일본
www.sony.com

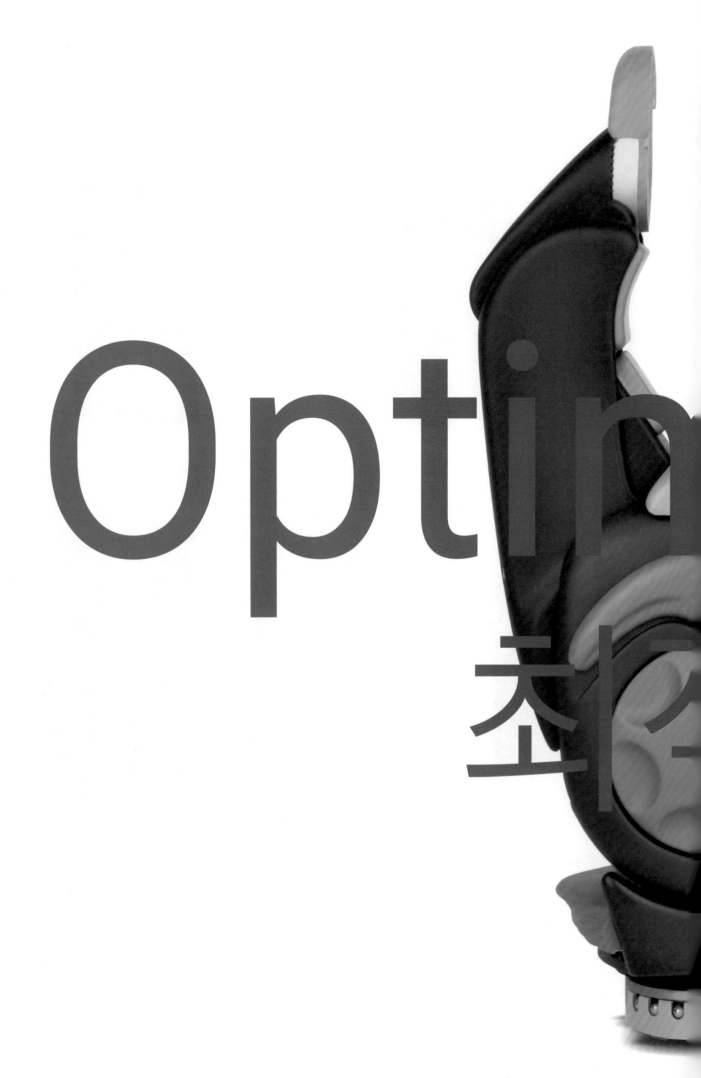

### System Electric Greifer DMC VariPlus

**Hand Prosthesis**
의수

**Manufacturer 제조**
Otto Bock Health Care
Products GmbH
빈 | 오스트리아
**In-house design 인하우스 디자인**
Annette Sting
www.ottobock.com
Page 114 | 115

### Xkelet

**Custom-Made Orthosis**
맞춤형 보조기기

**Manufacturer 제조**
Xkelet Easy Life S.L.
히로나 | 스페인
**In-house design 인하우스 디자인**
Andreu Carulla Studio
www.xkelet.com

# Optimising 최적화

The moment when human beings no longer accept their own biological weaknesses as simply natural or God-given, but instead optimise their bodies or senses by making use of the technological possibilities available to them, they enter into the realm of transhumanism. This is also where the boundaries between human being and machine become blurred, since artificial components become part of the natural organism and the artificial enters into a symbiosis with the natural. Design plays a big role here when it comes to the acceptance of prostheses or similar items. Either because such components are designed to be as natural as possible in appearance and motion so that they cannot easily be identified as artificial – or exactly the opposite, i.e. designed to intentionally emphasise the artificial. The latter ensures that the wearer presents his or her technologically optimised and occasionally even superior, artificial body part with self-confidence.

One example of this is the System Electric Greifer DMC VariPlus from Otto Bock: With its special arm prosthesis, it can be exchanged for a normal artificial hand and then offers precise gripping functions. With its reduced, strictly functional form, this special attachment calls to mind a high-quality tool – which it actually is. The addition of an LED lamp to illuminate the work area further underscores this impression. The gripper therefore enables the wearer to once again perform filigree activities and handle heavy equipment independently and confidently.

New technologies such as 3D printing also make it possible to manufacture medical prostheses and orthoses very individually and cost-effectively based on specific body data. The Xkelet, for instance, is an orthosis for immobilising damaged bones, which is made to measure by means of 3D printing. For this, the affected limb is scanned and measured. These data provide a 3D model that can be configured further using an app. An orthosis that is individually manufactured in this way offers greater comfort and mobility in comparison with rigid dressings and is also more hygienic; furthermore, it futuristic appearance delights.

인간은 더 이상 자신의 생물학적 약점을 그저 자연적인 것 또는 신께서 주신 것으로 만족하지 않고, 기술적 가능성을 활용하여 신체나 감각을 최적화하는 순간, 트랜스 휴머니즘의 영역으로 진입하게 된다. 이것은 인공적인 요소가 자연 유기체의 일부가 되고 인공적인 것이 자연적인 것과 공생하기 때문에 인간과 기계의 경계가 모호해지는 지점이기도 하다. 보철물 또는 이와 유사한 물건의 수용과 관련하여 디자인이 큰 역할을 하게 되는데, 그런 요소가 곧바로 인공물로 보이지 않도록 모양과 동작을 가능한 한 자연스럽게 디자인했거나, 아니면 정확히 그 반대, 즉 의도적으로 인공적인 면을 강조하도록 디자인되기 때문이다. 후자의 경우 기술적으로 최적화되고 때로는 더 우월한 인공 신체 부위를 착용하는 사람이 자신 있게 드러낼 수 있게 해준다.

이와 관련하여 Otto Bock의 System Electric Greifer DMC VariPlus를 예로 들 수 있다. 특별하게 제작된 의수는 일반 의수와 교체할 수 있으며 물건을 정확하게 잡는 기능을 제공한다. 정확히 의도된 최소한의 기능을 수행하도록 설계된 이 특별한 보철물은 고급 도구를 연상시킨다. 작업 공간을 밝히기 위해 LED 램프를 추가하면 이런 점이 더욱 강조된다. Greifer는 착용자가 섬세한 작업을 처리하고 무거운 장비를 혼자서 자신 있게 다룰 수 있게 해준다.

3D 프린팅과 같은 신기술도 특정 신체 데이터를 기반으로 개개인에 맞춘 의료용 보철물과 보조기를 비용 효율적으로 제작할 수 있게 해준다. 예를 들어 Xkelet은 손상된 뼈를 고정하기 위한 보조기이며 3D 프린팅으로 주문 제작된다. 이를 위해 착용 대상 팔다리를 스캔하고 측정한다. 이런 데이터는 앱을 사용하여 추가로 구성할 수 있는 3D 모델을 제공한다. 이렇게 개개인에 맞춰 제작된 보조기는 단단한 석고에 비해 사용하고 이동하기 편하며 더 위생적이다. 또한 미래지향적인 외관은 덤이다.

## Pudding BeanQ

**Robot for Early
Childhood Education**
유아교육용 로봇

**Manufacturer 제조**
Intelligent Steward Co., Ltd.
베이징 | 중국
**In-house design 인하우스 디자인**
Yi Chen, Feizi Ye, Tingting Xue,
Bin Zheng, Haichen Zheng,
Yong Zheng, Jian Sun, Ye Tian,
Xue Mei, Fan Li
www.roobo.com

## Interacting 교감

While, a few decades ago, robots still belonged to the realm of science fiction, they are today increasingly becoming part of our day-to-day life. Technology and Big Data have meanwhile become so advanced that the artificial creatures are able to take on ever more tasks. Robots are, so to say, the quintessence of this technological development. They function and operate on the basis of the gigantic amounts of data that digitalisation has created and, by means of constant feedback loops and machine learning, will be able to act ever more intelligently and independently in the not too distant future. Experts have long spoken about the greatest breakthrough since the Industrial Revolution. And just as at the beginning of the machine age, many people today regard the beginning of the robotic age with mixed feelings. The question that is therefore gaining in significance is how robots can be designed so that they are not seen as a competition or threat, but will instead accepted in future as a natural part of our lifestyle and working life.

수십 년 전만 해도 로봇은 공상과학 소설에서나 나오는 것이었지만 오늘날에는 점점 더 일상의 일부분이 되어가고 있다. 그 사이 기술과 빅 데이터가 발전하여 인공물이 훨씬 더 많은 일을 처리할 수 있게 되었다. 말하자면 로봇은 이런 기술 발전의 결정체이다. 로봇은 디지털화로 생성된 어마어마한 양의 데이터를 기반으로 작동하며, 지속적인 피드백과 머신 러닝을 통해 멀지 않은 미래에 더 지능적이고 독립적으로 작동할 수 있을 것이다. 전문가들은 산업혁명 이후 가장 위대한 발견에 대해 수년간 논의해 왔으며, 기계시대 초기와 마찬가지로 오늘날 많은 사람들은 로봇 시대의 서막을 복잡한 감정으로 맞이하고 있다. 그래서 점점 더 중요해지는 문제는 어떻게 하면 로봇을 경쟁상대나 위협이 아니라 향후 우리의 생활과 업무의 일부로 자연스럽게 받아들여지도록 디자인 하는가이다.

The general image of robots so far has been dominated by pop culture – by films, books, comics and manga – whereby what prevails is either the machine aspect or the humanoid, which focuses on the side of robots that is similar to human beings.

In the case of service or educational robots, hence robots that are found above all in the domestic and health care field, the focus is on the latter, i.e. at least on a distantly humanoid design. The reason for this is that personal robots are robots that assist people and are supposed to interact with them directly. To establish this kind of personal relationship, when designing robots and their human-machine-interface, manufacturers and designers focus primarily on two elements: language and eyes. Communication by means of a language-based user interface makes interaction with the machine easier and lends it a certain naturalness. Equipping a robot with eyes, on the other hand, enhances the ability to make contact, since a look from big, wide eyes winking or closing suggests liveliness and creates sympathy – even if the eyes are only a graphic representation on a display.

The small personal robot Pudding BeanQ is a successful example of the form that an intelligent, interactive robot can also take. The robot, which was designed especially for small children, is able to play with children, teach them, or play back video chats. It therefore represents a new design for a robot – its organic formal language calls a bean to mind. This kind of infantilised form dispels any impression of the robot being a threat and touches observers emotionally. The human-machine interaction also has a particularly child-friendly design. The large face area emphasises the focus on children and an extensive miming repertoire simplifies communication thanks to the wide range of emotions portrayed.

지금까지는 로봇에 대한 일반적인 이미지가 영화, 책, 만화 등 대중문화에 의해 소구되어 왔으며, 이런 대중문화에서는 기계적인 측면이나 인간과 비슷한 로봇의 측면에 초점을 맞춘 휴머노이드의 이야기가 지배적이다.

서비스, 교육용 로봇, 그리고 가정 및 의료 분야에서 볼 수 있는 로봇의 경우, 조금이라도 인간과 비슷한 디자인에 초점을 맞추고 있다. 그 이유는 개인용 로봇은 사람을 보조하고 사람과 직접 상호 작용하기 때문이다. 이런 개인적인 관계를 구축하기 위해 로봇의 인간-기계 인터페이스를 디자인할 때 제조업체와 디자이너는 주로 언어와 눈이라는 두 가지 요소에 중점을 둔다. 언어 기반 사용자 인터페이스를 통한 의사소통은 기계와의 상호작용을 더 쉽게 이루어지도록 하며 어느정도 자연스러운 느낌을 준다. 무엇보다 로봇에 눈을 장착하면 친밀감이 향상된다. 눈이 디스플레이 상의 그래픽 표현일지라도 크고 넓은 눈이 윙크를 하거나 눈을 감고 있는 모습이 생동감을 주며 공감을 불러일으키기 때문이다. 소형 개인 로봇 Pudding BeanQ는 지능형 대화형 로봇에도 적용할 수

있는 형태를 잘 보여주고 있다. 특히 어린 아이들을 위해 디자인된 이 로봇은 아이들과 놀거나 아이들을 가르치거나 화상 채팅을 재생할 수 있다. 로봇의 유기적인 조형 언어, 즉 눈동자의 모양은 '콩'을 떠올리게 하는 데, 이것이 바로 로봇 디자인에 있어 새로운 방향을 제시한다. 이렇게 아이들을 고려한 외관은 로봇이 위협적이라는 느낌을 없애고 보는 이에게 정서적으로 다가갈 수 있게 해준다. 이러한 로봇의 인간-기계 상호작용에도 어린이 친화적인 디자인이 적용되어 있다. 넓은 얼굴 부위는 어린이에게 초점을 맞춘 점을 강조하며, 여러 가지 표정은 다양하게 표현되는 감정을 통해 의사소통을 단순화시킨다.

## PuduBOT

**Service Robot**
서비스 로봇

**Manufacturer 제조**
Shenzhen Pudu
Technology Co., Ltd.
선전 | 중국
**In-house design 인하우스 디자인**
Peng Chen
www.pudutech.com

1

2

**1**

## iJINI

**Smart Robot**
스마트 로봇

**Manufacturer 제조**
nnovative Play Lab Co., LTD
고양 | 대한민국
**In-house design 인하우스 디자인**
Kyoung June Park
www.ipl.global

**2**

## Zenbo

**Home Robot**
가정용 로봇

**Manufacturer 제조**
ASUSTeK Computer Inc.
타이베이 | 대만
**In-house design 인하우스 디자인**
www.asus.com

**3**

## Midea Home

**Smart Voice Robot**
스마트 음성 로봇

**Manufacturer 제조**
Midea Smart Technology Co., Ltd.
선전 | 중국
**In-house design 인하우스 디자인**
Tao Xiao, Mingyu Xu, Ao Li
www.midea.com

**4**

## JELLY

**Commercial Service Robot**
상업용 서비스 로봇

**Manufacturer 제조**
Intelligent Steward Co., Ltd.
베이징 | 중국
**In-house design 인하우스 디자인**
Feizi Ye, Yong Zheng,
Haichen Zheng
www.roobo.com

3            4

# Delegating 위임

The arrival of the robot age raises the question of how much importance we want to give to autonomous machines in our private lives. In the field of household robots, the question is easy to answer. Their purpose consists mainly of relieving people by doing their work for them. For all practical purposes, such robots are therefore on the same level as washing machines or dishwashers, whose use is undisputed. In future, people will be able to delegate work such as vacuum cleaning, dusting, or mowing the lawn to an intelligent, mainly autonomously acting robot. These machines use sensors and GPS data about the space in which they are active and are hence able to move around independently. At present, many people are still torn between euphoria and scepticism regarding technology when it comes to the idea of allowing intelligent machines entry to their own rooms. One must nevertheless assume that – as with all great innovations – it is ultimately only a question of time before vacuum cleaning and dusting robots will be considered part of normal household equipment.

In the World Robot Report 2016, the International Federation of Robotics correspondingly predicted that the number of household robots will increase from 3.6 million devices worldwide in 2015 to 31 million units in 2019.

Interestingly, from a design perspective, household robots have absolutely nothing in common with the at least remotely humanoid robots propagated by pop culture. Their image is for the most part technically cool, objective, and characterised by basic geometrical forms. The main focus in the development and design of these household devices is their high degree of functionality. Thus, for example, the Smart Pro Compact vacuum-cleaner robot from Philips has a flat construction, while circularly arranged ribs on the cover express its high suction power.

로봇 시대의 도래는 우리가 사생활에서 자율적으로 동작하는 기계에 얼마나 큰 의미를 부여하고자 하는지를 묻는다. 가정용 로봇의 경우라면, 쉽게 대답할 수 있다. 가정용 로봇의 목적은 주로 사람들을 위해 일을 처리함으로써 사람들에게서 일에 대한 부담을 줄여주는 것이다. 따라서 모든 실용적인 목적을 위해 제작된 로봇들은 사용에 논란의 여지가 없는 세탁기 또는 식기 세척기와 같은 수준의 의미가 부여된다. 앞으로는 진공청소, 먼지털이, 또는 잔디 깎기와 같은 작업을 주로 자율적으로 작동하는 지능형 로봇에 맡길 수 있게 될 것이다. 이런 기계들은 센서와 GPS 데이터를 사용하여 자신이 작업하는 공간에서는 독립적으로 이동할 수 있다. 지능을 가진 기계가 자신의 방에 들어갈 수 있도록 하는 것과 관련하여 아직도 많은 사람들이 이런 기술에 대해 희열을 느끼는 사람과 회의적인 사람들로 갈라서 있다. 그러나 모든 위대한 혁신이 그랬던 것처럼 진공 청소 로봇과 먼지털이 로봇이 일반 가정에서 사용하는 기계장치로

취급되는 것은 결국 시간문제라고 보아야 한다.
국제로봇연맹(International Federation of Robotics)은 2016년 세계 로봇 보고서(World Robot Report 2016)에서 가정용 로봇의 수가 전세계적으로 2015년 360만 대에서 2019년 3100만 대로 증가할 것으로 예상했다.
디자인 관점에서 볼 때 흥미로운 점은 가정용 로봇이 대중문화에 의해 사람들에게 알려진 휴머노이드 로봇과 공통점이 전혀 없다는 점이다. 휴머노이드의 이미지는 대부분 기술적으로 멋져 보이고, 감정이 없고, 기본적으로 기하학적인 형태를 특징으로 한다. 이런 가정용 장치를 개발하고 디자인할 때 주로 중점을 두는 부분은 높은 기능성으로 볼 수 있는데, 필립스의 Smart Pro Compact 진공 청소 로봇을 예로 들면 평평한 구조로 되어있고 커버에 로고를 중심으로 파인 문양은 높은 흡입력을 표현하고 있다.

# Smart Pro Compact

**Robotic Vacuum Cleaner**
로봇청소기

**Manufacturer 제조**
Philips
베이징 | 중국
**In-house design** 인하우스 디자인
www.philips.com

**1**

**Da Zu Rui Shi Radar**

**Cleaning Robot**
청소 로봇

**Manufacturer 제조**
Shenzhen Han´s Lidar Technology
선전 | 중국
**Design 디자인**
Shenzhen Newplan Design Co., Ltd.
(Yin Chen, Jiajian Long)
선전 | 중국
www.hanslaser.com

**2**

**Kobold VR200**

**Robot Vacuum Cleaner**
로봇 청소기

**Manufacturer 제조**
Vorwerk Elektrowerke GmbH &
Co. KG
독일
**In-house design 인하우스 디자인**
Uwe Kemker, Vorwerk Design
부퍼탈 | 독일
**Design 디자인**
Phoenix Design GmbH & Co. KG
독일
www.vorwerk.com

**3**

**Roomba® 980**

**Vacuuming Robot**
로봇 청소기

**Manufacturer 제조**
iRobot
베드퍼드 | 메사추세츠 | 미국
**In-house design 인하우스 디자인**
www.irobot.com

**4**

**Winbot 950**

**Window Cleaning Robot**
유리창 청소 로봇

**Manufacturer 제조**
Ecovacs Robotics Co., Ltd.
쑤저우 | 중국
**In-house design 인하우스 디자인**
Li Xiaowen, Zhang Fan
www.ecovacs.com

**2**

**4**

# Collab 의

orating

## Collaborating 협업

Digitalisation has a big impact on the world of work and will change it fundamentally. Thanks to Big Data and networking, partially or completely autonomous machines are today able to manufacture products in a flexible and individualised manner. They take over monotonous or exhausting tasks or work that requires great precision. Programs based on neuronal networks and algorithms are also augmenting the artificial intelligence of machines, so that they are increasingly able to take on knowledge-related work as well. From a study in 2013, scientists of Oxford University deduced that, in twenty years, robots will perform at least forty-seven per cent of the work currently done in the United States. A driving force behind this is the desire to increase productivity and rationalisation. The replacement of human workers by robots is therefore propelled by strictly rational considerations. This makes the role played by the emotional quality in the design of these machines all the more important.

Designers counter the slight unease of persons who work side-by-side with such robots on a daily basis with a formal language that makes the robots seem less technoid and threatening. This can take the form of soft,

디지털화는 직업 세계에 큰 영향을 주며, 근무환경을 근본적으로 변화시킨다. 오늘날 빅데이터와 네트워킹으로 인하여 반자동 또는 완전자동 기계는 유연하고 개별화된 방식으로 제품을 생산할 수 있다. 이런 기계는 단조롭거나 노동 집약적이거나 높은 정밀도가 요구되는 작업에 투입된다. 신경망과 알고리즘을 기반으로 하는 프로그램도 기계가 지식과 관련된 업무에도 투입될 수 있도록 기계의 인공지능을 향상시키고 있다. 옥스포드 대학의 과학자들은 2013년에 진행한 연구에서, 현재 미국을 기준으로 20년 안에 로봇이 업무의 최소 47%를 수행하게 될 것으로 예측했다. 이에 대한 근거로 언급한 것은 생산성과 합리성을 높이고자 하는 열망이다. 다시 말해 인간 노동자를 로봇으로 대체하는 것은 순전히 합리적인 사고에 의한 것이기 때문에 이런 기계를 디자인할 때 감성품질이 맡은 역할이 더욱 중요해졌다. 디자이너는 디자인적 표현을 통해 로봇을 덜 딱딱하고 덜 위협적으로 보이게 함으로써, 매일 로봇과 나란히 일하는 사람들이 느낄 수 있는 일종의 불편함에 대응한다. 부드럽고 유기적인 선, 사람 같은 얼굴 또는 사람이나 동물을 연상시키는 다른 요소를 적용할 수 있다. 친숙한 모습과 작동순서를 적용하여 접촉에 대한 두려움

organic lines, a humanoid face, or another element that suggests a person or animal. It is about making use of familiar forms and movement sequences so as to dispel fear of contact and create acceptance. Industrial robots should not be perceived as a threat. They instead exist somewhere in the area of tension between artificial colleagues or assistants, perfect tools, and modern work slaves – according to how closely people and robots work together and how interaction between them is designed.

The small robot KR 3 Agilus manages the balancing act between highly efficient automated working aid, on the one hand, and emotional, pleasant appearance on the other. It is used in manufacturing and assembling the smallest components and is one of the fastest robots in its class. At the same time, with its organic formal language, human-like joint positions, and flowing contours and movements, the KR 3 Agilus seems lively and vaguely familiar – although it is anything but humanoid. Its form somehow remains abstract. It therefore shows how robots that work closely together with people can be designed without touching on the area of the so-called "uncanny valley", i.e. without seeming somehow threatening or weird.

## KR 3 AGILUS

**Small Robot**
소형 로봇

**Manufacturer** 제조
KUKA AG
아우크스부르크 | 독일
**In-house design** 인하우스 디자인
Christoph Groll,
Wolfgang Mayer
**Design** 디자인
Selic Industriedesign
아우크스부르크 | 독일
www.kuka.com

을 없애고 수용할 수 있게 하는 것이다. 산업용 로봇이 위협으로 인식되어서는 안된다. 인간과 로봇이 얼마나 밀접하게 협력하고 그들의 상호작용이 어떻게 의도되었는지에 따라 로봇 동료 또는 도우미, 완벽한 도구, 현대의 일꾼 사이의 무엇으로 정의된다.

소형 로봇 KR 3 Agilus는 고효율 자동 작업 보조장치이면서도 감성적이고 보기 좋은 모습으로 제작되었다. 가장 작은 부품을 생산하고 조립하는 데 사용되며 동급 로봇 중 가장 빠르다. 또한 KR 3 Agilus는 유기적인 조형 언어, 인간 같은 관절위치, 흐르는 듯한 윤곽과 동작 등, 인간형 로봇은 아니지만 생동감 있고 어느정도 친숙해 보이기도 한다. 왠지 추상적인 형태를 유지하고 있다고 볼 수 있다. 따라서 KR 3 Agilus는 사람들과 함께 가까이서 일하는 로봇을 '인간과 비슷해 보이는 로봇을 보면 생기는 불안감과 혐오감'을 피해 어떻게 든 위협적이거나 이상해 보이지 않게 디자인할 수 있는지를 보여주고 있다.

## Visualising 시각화

In the age of information, *Homo ex Data* has itself become a data provider. Its body data flow into worldwide algorithms, feed the data flow, and hence facilitate ever more exact diagnoses and more personalised treatment. High-tech and medicine are closely linked and complement each other seamlessly. Human beings are still extremely complex organisms, but, in the past decades, they have nevertheless become easier and easier to analyse and have, in principle, consequently become manageable systems. Data from inside the body make a great contribution to this. Such data include not only blood analyses with their values, electrocardiograms, electroencephalography, and similar procedures, but also imaging techniques such as sonography, X-rays, and computer tomography. With the aid of such procedures, human beings are illuminated – the invisible is made visible. This makes it possible to identify abnormalities more easily and to analyse and classify values and visualisations thanks to a continuously growing pool of comparative data.

Thanks to big advances in the processing of digital signals and greater computer efficiency, ultrasound devices, for example, today offer new possible applications. Ultrasound devices compute images from reflected sound waves. Noise interference can be suppressed or eliminated by means of image optimising procedures such as sound wave encoding, which means that images today are considerably better, and even three- or four-dimensional photos and panorama images from the inside of the body can be provided and then saved and digitally enhanced. One example of this is the AX8 ultrasound system in laptop format, which generates high-quality diagnostic images for human and veterinary medicine. It incorporates a sealed user interface with two touch panels, elastomeric keys, and gesture control, thus making it possible for users to concentrate entirely on the patient and the images.

정보화 시대에 호모엑스 데이터는 그 자체가 데이터 제공자가 되었다. 신체 데이터는 전 세계 알고리즘으로 흘러들어 데이터 흐름을 공급하며더욱 정확한 진단과 개인화된 치료가 용이하게 되었다. 첨단기술과 의학은 긴밀하게 연결되어 있고, 서로 완벽하게 보완한다. 인간은 여전히 극도로 복잡한 유기체이지만, 그럼에도 불구하고 지난 수십 년 동안 분석하기가 더 쉽고, 원칙적으로 관리 가능한 시스템이 되었다. 신체 내부에 대한 데이터가 이에 크게 기여하고 있다고 볼 수 있는데, 혈액 분석, 심전도 검사, 뇌파검사, 및 이와 유사한 검사, 그리고 초음파촬영, 엑스레이, 컴퓨터 단층촬영과 같은 영상 기술이 포함된다. 이런 검사와 기술을 통해 보이지 않던 인체 내부를 볼 수 있게 되었다. 이로 인해 이상한 점을 보다 쉽게 파악할 수 있게 되었고 비교 데이터가 지속적으로 증가하여 측정값을 분석 및 분류하고 시각화 할 수 있게 되었다.

디지털 신호 처리기술의 커다란 진보와 컴퓨터 효율의 향상에 힘입어 오늘날의 초음파 장치는 새로운 응용분야를 찾게 되었다. 초음파 장치는 반사된 음파로부터 이미지를 계산한다. 음파 인코딩과 같은 이미지 최적화를 통해 노이즈 간섭을 억제하거나 제거할 수 있어 오늘날에는 영상이 상당히 개선되었고 신체 내부를 3차원 또는 4차원 사진, 그리고 파노라마 영상으로도 보여줄 수 있게 되었으며, 저장하여 디지털 기술로 향상시킬 수도 있게 되었다. 랩톱 형식의 AX8 초음파 시스템은 인체용, 수의학용 고품질 진단 영상을 생성하며, 밀봉된 사용자 인터페이스를 두 개의 터치 패널, 엘라스토머 키, 제스처 제어장치와 통합시켜 사용자가 환자와 영상에 완전히 집중할 수 있도록 한다.

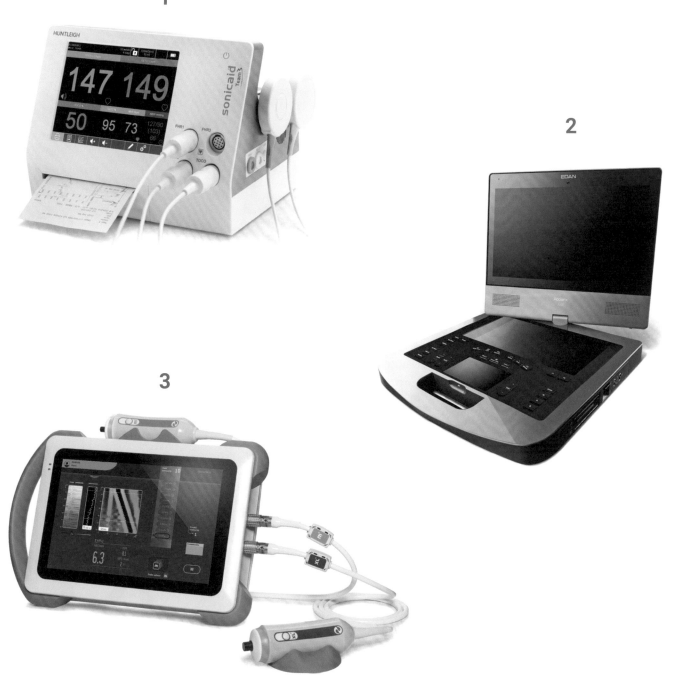

**1**

# Sonicaid Team 3

**Fetal Monitor**
태아 심박수 모니터

**Manufacturer 제조**
Huntleigh
카디프 | 영국
**Design 디자인**
PDR, Cardiff, Great Britain
www.huntleigh-diagnostics.com

**2**

# AX8

**Laptop Ultrasound System**
노트북 초음파 시스템

**Manufacturer 제조**
Edan Instruments, Inc.
선전 | 중국
**In-house design 인하우스 디자인**
Richard Henderson
www.edan.com.cn

**3**

# FibroScan Mini 430

**Device for Non-Invasive
Liver Diagnosis**
비침습적 간 진단 장치

**Manufacturer 제조**
Echosens
파리 | 프랑스
**Design 디자인**
Nova Design (Olivier Jeanjean)
뤼넬 | 프랑스
www.echosens.com

# Experi

체

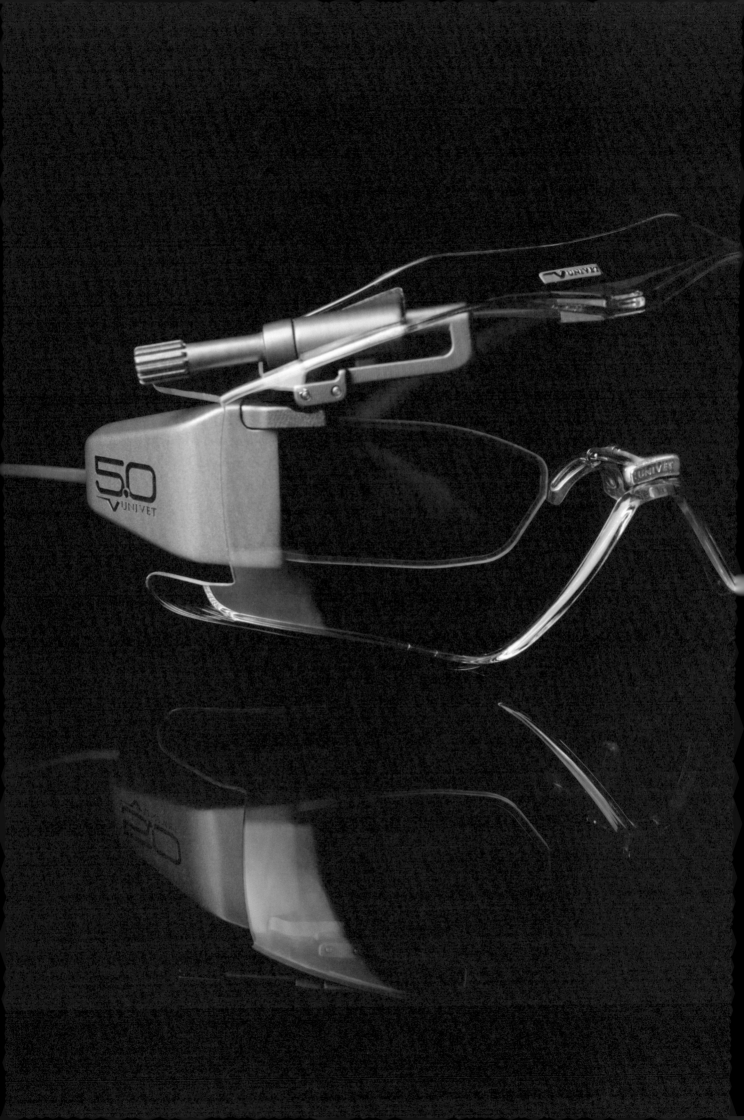

# Experiencing 체험

Digitalisation and networking have now changed and simplified peoples lives to such a great extent and simultaneously made them so much more complex that they differ fundamentally from the lives that people lived only a couple of decades ago. Online, communication channels between people are short, even if they live thousands of kilometres apart in the analogue world. *Homo ex Data* is potentially networked with anyone who has access to the internet. Everyone also has access to enormous data and, therefore, knowledge resources. Digitalisation consequently contributes to extending our individual living environment. In addition, the most diverse information technologies are currently being further developed, with the result that the boundaries between the analogue and the virtual world are becoming ever more porous. Whether computer-simulated reality, a mix of reality and computer-generated content, or augmented reality – with virtual, mixed, and augmented reality, new, complex forms of reality that will significantly shape and expand people's world of experience in the years to come are coming into being. These are innovative types of media with enormous potential – even if many of these technologies in the field of virtual reality

## Univet 5.0

**Safety Glasses with Augmented Reality**
증강 현실 보안경

**Manufacturer 제조**
Univet
브레시아 | 이탈리아
**In-house design 인하우스 디자인**
Fabio Borsani
www.univet-optic.com

디지털화와 네트워킹으로 인해 이제 사람들의 삶이 크게 변하고 단순화되었으며 동시에 훨씬 더 복잡해졌다. 불과 수십 년 전 살았던 사람들의 삶과는 근본적으로 달라지게 되었다. 아날로그 세계에서는 수천 킬로미터 떨어져 있는 사람들이라도 온라인 통신을 통해 더 가깝게 느낄 수 있다. 호모 엑스 데이터는 인터넷에 액세스할 수 있는 모든 사람과 잠재적으로 네트워크를 통해 연결되어 있다. 또한 모든 사람은 방대한 양의 데이터에 액세스할 수 있으므로 지식의 자원에 액세스할 수 있다. 결과적으로 디지털화는 개인의 생활환경을 확장하는 데 기여한다. 또한, 현재 다양한 정보 기술이 더욱 발전하고 있으며, 그 결과 아날로그와 가상세계의 경계가 점점 허물어지고 있다. 컴퓨터 시뮬레이션 현실이든, 현실과 컴퓨터로 생성된 콘텐츠의 혼합이든, 증강현실이든, 가상, 혼합, 증강 현실과 함께 앞으로 몇 년 동안 사람들의 경험세계를 크게 형성하고 확장할 새롭고 복잡한 형태의 현실이 생겨나고 있다. 이런 가상현실 분야 기술 중 대부분이 아직 대중에게 노출되지는 않았지만, 이들은 엄청난 잠재력을 지닌 혁신적인 유형의 미디어이다. 그래도 미래에는 이런 세상들

have not yet become available to the masses. Nevertheless, in future, linking these worlds will have a great influence on the way in which we consume media, play, work, shop, and communicate.

In order to be able to become completely immersed in a virtual world based on computer simulation, whereby people can move relatively freely, touch things, or meet other people, aids such as the virtual reality spectacles 360 VR from LG are required. In addition, products such as the Exoskelett Dexmo for VR applications translate the wearer's real hand and finger gestures for the digital world and simultaneously provide haptic feedback in real time.

Products for the field of augmented reality are already considerably more widespread, as is shown by the example of head-up displays for motor vehicles. The Navdy uses augmented reality technology to project maps, phone calls, news, and flight information directly onto drivers' fields of vision, without restricting their view. The Microsoft HoloLens augmented reality glasses do not blend out reality, but instead enable holographic objects to enter it. Interaction with these objects takes place by means of looks, the voice, or gestures.

을 연결하는 것이 우리가 미디어를 소비하고, 즐기고, 일하고, 쇼핑하고, 소통하는 방식에 큰 영향을 미칠 것이다.

사람들이 비교적 자유롭게 움직이고, 물건을 만지고, 다른 사람을 만날 수 있는 컴퓨터 시뮬레이션 기반 가상세계에 완전히 몰입할 수 있으려면 LG의 가상현실 글래스 360 VR과 같은 보조 장치가 필요하다. 또한 VR 어플리케이션용 Exoskelett Dexmo와 같은 제품은 디지털 세계에서 착용자의 손과 손가락의 실제 제스처를 입력하고 동시에 실시간으로 촉각적인 피드백을 제공한다.

차량용 전방표시장치처럼 증강현실 분야의 제품도 이미 매우 널리 보급되어 있다. Navdy는 운전 시 시야를 방해하지 않으면서도 증강현실 기술을 사용하여 지도, 통화, 뉴스, 비행 정보를 운전자의 시야 내에서 직접 표시해준다. Microsoft HoloLens 증강현실 글래스는 현실을 투사하지 않고 대신 홀로그램 개체를 현실에 투입한다. 사용자는 이런 외형, 음성, 제스처를 통해 개체와 상호작용하게 된다.

## Microsoft HoloLens

**Augmented Reality Headset**
증강 현실 헤드셋

**Manufacturer 제조**
Microsoft Corporation
레드먼드 | 워싱턴 | 미국
**In-house design 인하우스 디자인**
www.microsoft.com/surface

**1**

**4**

**5**

## 1

### Dexmo

**Exoskeleton for Virtual Reality Applications**
가상 현실 전용 기기

**Manufacturer 제조**
Dexta Robotics Inc., Elko
엘코 | 미국 / 선전 | 중국
**Design 디자인**
Hefei XIVO Design Co., Ltd.
(Wenbao Chu, Chao Jing)
허페이 | 중국
www.DEXTAROBOTICS.com

## 2

### HiAR Glasses

**AR Glasses**
AR 안경

**Manufacturer 제조**
HiScene Information
Technology Co., Ltd.
상하이 | 중국
**Design 디자인**
LKK Design Shanghai Co., Ltd.
상하이 | 중국
www.hiscene.com

## 3

### 360 VR

**Head-Mounted Display**
헤드 마운트 디스플레이

**Manufacturer 제조**
LG Electronics Electronics Inc.
서울 | 대한민국
**In-house design 인하우스 디자인**
Dongsoon Kim, Hyunchul Kim,
Seungyup Lee
www.lg.com

## 4

### Navdy

**Head-up Display**
헤드 업 디스플레이

**Manufacturer 제조**
Navdy
샌프란시스코 | 캘리포니아 | 미국
**In-house design 인하우스 디자인**
Jesse Madsen
www.navdy.com

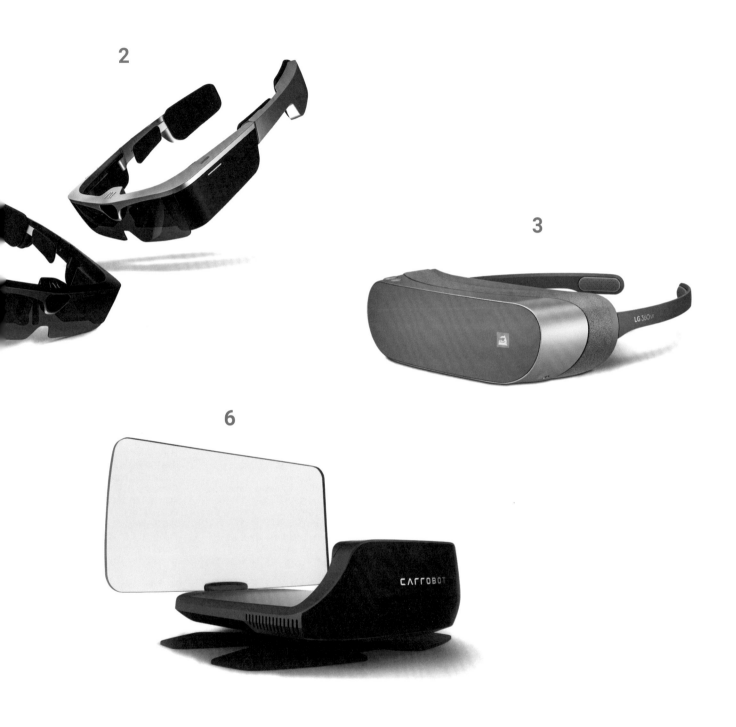

2

3

6

**5**
## MovieMask

**VR Headset**
VR 헤드셋

**Manufacturer 제조**
MovieMask AS
트론헤임 | 노르웨이
**Design 디자인**
Inventas AS
트론헤임 | 노르웨이
www.moviemask.io

**6**
## Carrobot C2

**Head-up Display**
헤드 업 디스플레이

**Manufacturer 제조**
Beijing iLeja Tech Co., Ltd.
베이징 | 중국
**Design 디자인**
LKK Design Beijing Co., Ltd.
베이징 | 중국
www.carrobot.com

# Index 찾아보기

# D

Da Zu Rui Shi Radar
Cleaning Robot
청소 로봇 2016년
Shenzhen Han´s Lidar Technology
Shenzhen | China 선전 | 중국
Design 디자인
Shenzhen Newplan Design
Shenzhen | China 선전 | 중국

Deebot DT85
Cleaning Robot
청소 로봇 2016년
Ecovacs Robotics
Suzhou | China 쑤저우 | 중국
In-house design 인하우스 디자인

DeeBbot Slim 3s
Cleaning Robot
청소 로봇 2016년
Ecovacs Robotics
Suzhou | China 쑤저우 | 중국
In-house design 인하우스 디자인

Dell Latitude 2-in-1
Two-in-One Notebook
투인원 노트북 2017년
Dell
Round Rock | TX | USA
라운드 록 | 텍사스 | 미국
In-house design 인하우스 디자인

Deluxe Bus Video Panel
BVPC 850
Bus-Video-Panel
Comfort BVPC 850
버스 비디오 패널 2013년
S. Siedle & Söhne
Furtwangen | Germany 푸트반겐 | 독일
In-house design 인하우스 디자인

DeWalt Rotary Laser
Rotary Laser Layout Device
회전 레이저 2016년
Stanley Black & Decker
Southington | CT | USA
사우딩턴 | 코네티컷 | 미국
In-house design 인하우스 디자인

Dexmo
Exoskeleton for Virtual Reality
Applications
가상 현실 전용 외골격 2017년
Dexta Robotics Dexta
Elko | USA 엘코 | 미국
Design 디자인
Hefei XIVO Design
Hefei | China 허페이 | 중국

Dräger Babyleo® TN500
Incuwarmer
인큐베이터 2016년
Drägerwerk
Lübeck | Germany 뤼베크 | 독일
Design 디자인
MMID
Essen | Germany 에센 | 독일

DS2
Rugged Handheld Computer
산업용 핸드헬드 컴퓨터 2017년
DSIC
Seoul | South Korea 서울 | 대한민국
Design 디자인
Design NID
Seoul | South Korea 서울 | 대한민국

# E

Eargo
Hearing Aid
보청기 2015년
Eargo
Mountain View | CA | USA
마운틴 뷰 | 캘리포니아 | 미국
In-house design 인하우스 디자인

EC-760ZP-V/L
Video Endoscope
비디오 내시경 2016년
Fujifilm
Tokyo | Japan 도쿄 | 일본
In-house design 인하우스 디자인

EF500 Series
Smartphone
스마트폰 2016년
Bluebird
Seoul | South Korea 서울 | 대한민국
In-house design 인하우스 디자인

Engine 27
CPU Cooler
CPU 쿨러 2016년
Thermaltake Technology
Taipei | Taiwan 타이베이 | 대만
In-house design 인하우스 디자인

Enobio 2 StarStim
Brain-Monitoring Helmet
뇌 모니터링 헬멧 2015년
Neuroelectrics
Barcelona | Spain 바르셀로나 | 스페인
Design 디자인
Ànima design
Barcelona | Spain 바르셀로나 | 스페인

# F

FibroScan Mini 430
Device for Non-Invasive
Liver Diagnosis
비침습적 간 진단 장치 2016년
Echosens
Paris | France 파리 | 프랑스
Echsens
Design 디자인
Nova Design
Lunel | France 뤼넬 | 프랑스

Finepix XP90
Digital Camera
디지털카메라 2015년
Fujifilm
Tokyo | Japan 도쿄 | 일본
In-house design 인하우스 디자인

FoldIT® USB
USB Flash Drive
USB 플래쉬 드라이브 2016년
CustomUSB
Wheeling | IL | USA
휠링 | 일리노이 | 미국
Design 디자인
ClevX
Kirkland | WA | USA
커클랜드 | 워싱턴 | 미국

Foream Compass
Lifestyle Video Camera
라이프스타일 비디오 카메라 2016년
Foream
Shenzhen | China 선전 | 중국
In-house design 인하우스 디자인

# G

GIS 1000 C Professional
Thermo Detector
열 감지기 2015년
Robert Bosch Elektrowerkzeuge
Leinfelden-Echterdingen | Germany
라인펠덴에히터딩엔 | 독일
In-house design 인하우스 디자인

GoPro Karma System
Drone
드론 2017년
GoPro
San Mateo | CA | USA
샌머테이오 | 캘리포니아 | 미국
In-house design 인하우스 디자인

# H

HiAR Glasses
AR Glasses
AR 안경 2017년
HiScene Information Technology
Shanghai | China 상하이 | 중국
Design 디자인
LKK Design
Shanghai | China 상하이 | 중국

Hover Camera
Drone
드론 2016년
Zero Zero Robotics
Beijing | China 베이징 | 중국
In-house design 인하우스 디자인

HP Elite Slice
Computer
컴퓨터 2016년
HPPalo Alto | CA | USA
팔로 알토 | 캘리포니아 | 미국
In-house design 인하우스 디자인

HP Envy Phoenix
Computer
컴퓨터 2015년
HP
Palo Alto | CA | USA
팔로 알토 | 캘리포니아 | 미국
In-house design 인하우스 디자인

MaxiSys® MS906
Automotive Diagnostic Device
차량 진단장치 2016년
Autel Intelligent Technology
Shenzhen | China
선전 | 중국
In-house design 인하우스 디자인

MaxShot 3D
Optical Measuring System
광학 측정 시스템 2017년
Creaform
Lévis | Canada 퀘백 | 캐나다
In-house design 인하우스 디자인

MetraScan 3D / HandyProbe /
C-Track
Portable 3D Measurement System
휴대용 3D 측정 시스템 2016년
Creaform
Lévis | Canada 퀘백 | 캐나다
In-house design 인하우스 디자인

Mi Powerline wifi
Powerline Adapter
파워라인 어댑터 2017년
Beijing Xiaomi Mobile Software
Beijing | China 베이징 | 중국
In-house design 인하우스 디자인

Mi Router 3 Family
Routers
라우터 2016년
Beijing Xiaomi Mobile Software
Beijing | China 베이징 | 중국
In-house design 인하우스 디자인

Mi Router Pro Family
Routers
라우터 2017년
Beijing Xiaomi Mobile Software
Beijing | China 베이징 | 중국
In-house design 인하우스 디자인

Microsoft HoloLens
Augmented Reality Headset
증강 현실 헤드셋 2016년
Microsoft
Redmond | WA | USA
레드먼드 | 워싱턴 | 미국
In-house design 인하우스 디자인

Midea Bubble Robot
Intelligent Voice Robot
지능형 음성 로봇 2017년
Midea Smart Technology
Shenzhen | China 선전 | 중국
In-house design 인하우스 디자인

Midea Home
Smart Voice Robot
스마트 음성 로봇 2017년
Midea Smart Technology
Shenzhen | China 선전 | 중국
In-house design 인하우스 디자인

Midea o2
Smart Home Management
System
스마트 홈 제어 시스템 2017년
Midea Smart Technology
Shenzhen | China 선전 | 중국
In-house design 인하우스 디자인

Mini POS
POS System
POS 시스템 2016년
Taiwan Compal Electronics
Company
Taoyuan | Taiwan 타오위안 | 대만
Design 디자인
Avalue Technology
New Taipei City | Taiwan 신베이 | 대만

Mio Slice Band
Heart Rate and Health Tracker
심박수 건강 트랙커 2017년
Mio Global
Vancouver | Canada 밴쿠버 | 캐나다
Design 디자인
Woke Studio
Vancouver | Canada 밴쿠버 | 캐나다

MONO MO-01J
Smartphone
스마트폰 2016년
NTT Docomo
Tokyo | Japan 도쿄 | 일본
In-house design 인하우스 디자인

MOOV NOW™
Sports Tracker
스포츠 트랙커 2015년
Moov
Burlingame | CA | USA
벌링게임 | 캘리포니아 | 미국
In-house design 인하우스 디자인

Motiv Ring
Fitness Tracker
피트니스 트랙커 2017년
Motiv
San Francisco | CA | USA
샌프란시스코 | 캘리포니아 | 미국
In-house design 인하우스 디자인

Motorola AC1700
Router
라우터 2017년
Shenzhen Gongjin Electronics
Shenzhen | China 선전 | 중국
In-house design 인하우스 디자인

MovieMask
VR Headset
VR 헤드셋 2016년
MovieMask
Oslo | Norway 오슬로 | 노르웨이
Design 디자인
Inventas
Trondheim | Norway
트론헤임 | 노르웨이

MyAir
Portable Air Sensor
휴대용 공기 센서 2017년
Tion
Novosibirsk | Russia
노보시비르스크 | 러시아
Design 디자인
Logeeks
Novosibirsk | Russia
노보시비르스크 | 러시아

N

N60NC Wireless
Wireless Headphones
무선 헤드폰 2017년
Harman International Industries
Northridge | CA | USA
노스리지 | 캘리포니아 | 미국
In-house design 인하우스 디자인

Navdy
Head-up Display
헤드 업 디스플레이 2016년
Navdy
San Francisco | CA | USA
샌프란시스코 | 캘리포니아 | 미국
In-house design 인하우스 디자인

Nighthawk X10
Router
라우터 2016년
Netgear
San Jose | CA | USA
산호세 | 캘리포니아 | 미국
In-house design 인하우스 디자인

Nvidia Geforce GTX 1060
Graphics Card
그래픽 카드 2016년
Nvidia
Santa Clara | CA | USA
산타클라라 | 캘리포니아 | 미국
In-house design 인하우스 디자인

Nvidia Geforce GTX 1080
Graphics Card
그래픽 카드 2016년
Nvidia
Santa Clara | CA | USA
산타클라라 | 캘리포니아 | 미국
In-house design 인하우스 디자인

Nvidia Shield Android TV
TV Box Set
TV 박스 세트 2015년
Nvidia
Santa Clara | CA | USA
산타클라라 | 캘리포니아 | 미국
In-house design 인하우스 디자인

O

O6
Intelligent Smartphone Controller
스마트폰 컨트롤러 2017년
Fingertips Lab
San Jose | CA | USA
산호세 | 캘리포니아 | 미국
In-house design 인하우스 디자인

OCLU
Action Camera
액션 카메라 2017년
OCLU
London | Great Britain 런던 | 영국
In-house design 인하우스 디자인

Ozo
Virtual Reality Camera
VR 카메라 2016년
Nokia Technologies
Sunnyvale | CA | USA
서니베일 | 캘리포니아 | 미국
In-house design 인하우스 디자인

P

Paxton Entry Touch
Door Intercom Panel
현관 인터콤 패널 2017년
Paxton Access
Brighton | Great Britain
브라이턴 | 영국
In-house design 인하우스 디자인

PEN-F
System Camera
시스템 카메라 2016년
Olympus Europa
Hamburg | Germany 함부르크 | 독일
Design 디자인
Olympus Corporation
Design Center
Tokyo | Japan 도쿄 | 일본

Phantom 4 Pro
Drone
드론 2016년
DJI
Shenzhen | China 선전 | 중국
In-house design 인하우스 디자인

Phonak Audéo™ V10
Hearing Aid
보청기 2014년
Sonova
Stäfa | Switzerland 슈테파 | 스위스
In-house design 인하우스 디자인

Play Collection
Wireless Mouse
무선 마우스 2015년
Logitech
Newark | CA | USA
뉴어크 | 캘리포니아 | 미국
In-house design 인하우스 디자인

Power Mac G5
Desktop computer
데스크탑 컴퓨터 2003년
Apple
Cupertino | CA | USA
쿠퍼티노 | 캘리포니아 | 미국
In-house design 인하우스 디자인

PowerEgg
Quadrocopter
쿼드콥터 2016년
PowerVision Robot
Beijing | China 베이징 | 중국
In-house design 인하우스 디자인

PowerEye
Quadrocopter
쿼드콥터 2016년
PowerVision Technology
Beijing | China 베이징 | 중국
In-house design 인하우스 디자인

Pudding BeanQ
Robot for Early Childhood
Education
유아교육용 로봇 2017년
Intelligent Steward
Beijing | China 베이징 | 중국
In-house design 인하우스 디자인

PuduBot
Service Robot
서비스 로봇 2017년
Shenzhen Pudu Technology
Shenzhen | China 선전 | 중국
In-house design 인하우스 디자인

PX7
POS Terminal
POS 단말기 2015년
PAX Computer Technology
Shenzhen | China 선전 | 중국
Design 디자인
Shenzhen ND Industrial Design
Shenzhen | China 선전 | 중국

# Q

Q-tube Wi-Fi Teeth Scope Pro
Teeth Scope
치아 스코프 2017년
Quanta Computer
Taoyuan | Taiwan 타오위안 | 대만
In-house design 인하우스 디자인

Q-tube Wi-Fi Otoscope Pro
ENT Otoscope
이비인후과 검이경 2017년
Quanta Computer
Taoyuan | Taiwan 타오위안 | 대만
In-house design 인하우스 디자인

# R

Raiden Series
Gaming Memory Modules
게임 메모리 모듈 2015년
AVEXIR Technologies
Zhubei | Taiwan 주베이 | 대만
In-house design 인하우스 디자인

RD TCL RC800 Series
Remote Controls
리모콘 2016년
TCL Multimedia Technology
Shenzhen | China 선전 | 중국
In-house design 인하우스 디자인

Robin
Smartphone
스마트폰 2015년
Nextbit
San Francisco | CA | USA
샌프란시스코 | 캘리포니아 | 미국
In-house design 인하우스 디자인

ROG Maximus IX Apex
Motherboard
마더보드 2017년
Asustek Computer
Taipei | Taiwan 타이베이 | 대만
In-house design 인하우스 디자인

ROG Maximus IX Extreme
Motherboard
마더보드 2017
Asustek Computer
Taipei | Taiwan 타이베이 | 대만
In-house design 인하우스 디자인

Rolly Keyboard 2
Bluetooth Keyboard
블루투스 키보드 2016년
LG Electronics
Seoul | South Korea 서울 | 대한민국
In-house design 인하우스 디자인

Roomba® 980
Vacuuming Robot
진공청소기 로봇 2015년
iRobot
Bedford | MA | USA
베드퍼드 | 메사추세츠 | 미국
In-house design 인하우스 디자인

# S

SD700
External SSD
외장 SSD 2016년
Adata Technology
New Taipei City | Taiwan 신베이 | 대만
In-house design 인하우스 디자인

Smart Carbon
Monoxide Sensor
CO-Detector
일산화탄소 감지기 2016년
Shenzhen Heiman Technology
Shenzhen | China 선전 | 중국
In-house design 인하우스 디자인

Smart Commander
Decoding Network Keyboard
네트워크 디코딩 키보드 2017년
Zhejiang Dahua Technology
Hangzhou | China 항저우 | 중국
In-house design 인하우스 디자인

Smart Pro Compact
Robotic Vacuum Cleaner
로봇청소기 2016년
Philips
Eindhoven | Netherlands
에인트호번 | 네덜란드
In-house design 인하우스 디자인

Soil Scanner
토양 스캐너 2017년
SoilCares
Wageningen | Netherlands
바헤닝언 | 네덜란드
Design 디자인
Scope Design & Strategy
Amersfoort | Netherlands
아메르스포르트 | 네덜란드

Sonicaid Team 3
Fetal Monitor
태아 모니터 2016년
Design PDR
Cardiff | Great Britain 카디프 | 영국
Design 디자인
Huntleigh
Cardiff | Great Britain 카디프 | 영국

SP-900, CP-900, RS-900
Endoscopic Ultrasound
Probe System
내시경 초음파 시스템 2016년
Fujifilm
Tokyo | Japan 도쿄 | 일본
In-house design 인하우스 디자인

STU-540
Signature Pad
서명 패드 2016년
Wacom
Tokyo | Japan 도쿄 | 일본
In-house design 인하우스 디자인

Swisscom TV 2.0
Remote Control
리모컨 2016년
Swisscom
Zurich | Switzerland 취리히 | 스위스
Design 디자인
ruwido austria
Neumarkt | Austria
노이마르크트 | 오스트리아

System Electric Greifer DMC VariPlus
Hand Prosthesis
의수 2017년
Otto Bock Health Care Products
Vienna | Austria 빈 | 오스트리아
In-house design 인하우스 디자인

# T

TeraStation TS Series
NAS Device
NAS 장치 2016년
Buffalo
Nagoya | Japan 나고야 | 일본
In-house design 인하우스 디자인

ThinkCentre M910 Tower
Computer
컴퓨터 2017년
Lenovo
Morrisville | NC | USA
모리스빌 | 노스캐롤라이나 | 미국
In-house design 인하우스 디자인

ThinkPad X1 Yoga
Convertible Notebook
컨버터블 노트북 2017년
Lenovo
Morrisville | NC | USA
모리스빌 | 노스캐롤라이나 | 미국
In-house design 인하우스 디자인

Toughpower DPS G RGB
PC Power Supply Unit
PC 전원 공급 장치 2016년
Thermaltake Technology
Taipei | Taiwan 타이베이 | 대만
In-house design 인하우스 디자인

TRX Connected Transportation Plat-
form
Rail Communication System
철도 통신 시스템 2017년
Klas Telecom
Dublin | Ireland 더블린 | 아일랜드
In-house design 인하우스 디자인

# U

Univet 5.0
Safety Glasses with Augmented
Reality
증강 현실 보안경 2017년
Univet
Rezzato | Italy 브레시아 | 이탈리아
In-house design 인하우스 디자인

# V

Vivideo ENT
Video Endoscope
이비인후과 비디오 내시경 2014년
Pentax Medical
Friedberg | Germany
프리트베르크 | 독일
In-house design 인하우스 디자인

vivomove Asia edition
Fitness Tracker
피트니스 트랙커 2016년
Garmin
New Taipei City | Taiwan 신베이 | 대만
In-house design 인하우스 디자인

Vodafone GigaTV
Remote Control
리모콘 2017년
Vodafone Kabel Deutschland
Unterföhring | Germany 운터퍼링 | 독일
Design 디자인
Thorsten Buch, Christian Olufemi
Munich | Germany 뮌헨 | 독일

# W

Cintiq 27QHD touch
Pen Display
펜 디스플레이 2015년
Wacom
Tokyo | Japan 도쿄 | 일본
Design 디자인
dingfest | design
Erkrath | Germany 에어크라트 | 독일

Walkman NW-WM1Z
Portable Audio Player
휴대용 오디오 플레이어 2016년
Sony Video & Sound Products
Tokyo | Japan 도쿄 | 일본
Design 디자인
Sony
Tokyo | Japan 도쿄 | 일본

Winbot 950
Window Cleaning Robot
유리창 청소 로봇 2016년
Ecovacs Robotics
Suzhou | China 쑤저우 | 중국
In-house design 인하우스 디자인

WT6000, RS6000
Industrial Wearable Computing
System
산업용 웨어러블 컴퓨팅 시스템 2016년
Zebra Technologies
Holtsville | NY | USA 홀츠빌 | 뉴욕 | 미국
In-house design 인하우스 디자인

# X

Xbox Elite
Wireless Controller
무선 컨트롤러 2016년
Microsoft
Redmond | WA | USA
레이먼드 | 워싱턴 | 미국
In-house design 인하우스 디자인

Xkelet
Custom-Made Orthosis
맞춤형 보조기 2016년
Xkelet Easy Life
Girona | Spain 히로나 | 스페인
In-house design 인하우스 디자인

Xperia X Compact
Smartphone
스마트폰 2016년
Sony Mobile Communications
Tokyo | Japan 도쿄 | 일본
Design 디자인
Sony
Tokyo | Japan 도쿄 | 일본

Xperia XA
Smartphone
스마트폰 2016년
Sony Mobile Communications
Tokyo | Japan 도쿄 | 일본
Design 디자인
Sony
Tokyo | Japan 도쿄 | 일본

Xplorer
Drone
드론 2015년
Shenzhen Rapoo Technology
Shenzhen | China 선전 | 중국
Design 디자인
Shenzhen zero-tech UAV
Shenzhen | China 선전 | 중국

Xplorer X2
Drone
드론 2016년
Shenzhen Rapoo Technology
Shenzhen | China 선전 | 중국
Design 디자인
Shenzhen Zero-Tech UAV
Shenzhen | China 선전 | 중국

XQ2
Digital Camera
디지털카메라 2015년
Fujifilm
Tokyo | Japan 도쿄 | 일본
In-house design 인하우스 디자인

XS
Drone
드론 2016년
Shenzhen Rapoo Technology
Shenzhen | China 선전 | 중국
Design 디자인
Shenzhen Zero-Tech UAV
Shenzhen | China 선전 | 중국

# Z

Zenbo
Home Robot
가정용 로봇 2017년
Asustek Computer
Taipei | Taiwan 타이베이 | 대만
In-house design 인하우스 디자인

# Imprint

**Idea and Concept**
Peter Zec

**Project Supervision**
Vito Oražem

**Project Assistance**
Gretha Lösch-Schloms

# Homo Ex Data

## The Natural of the Artificial

# Publication

**Editors**
Burkhard Jacob
Vito Oražem
Peter Zec

**Editorial Work**
Astrid Ruta, Konzept | Text | Redaktion
Essen | Germany

**Texts**
Astrid Ruta
Peter Zec

**Design Concept**
zinnobergruen
Düsseldorf | Germany

**Layout**
Maren Reinecke
Goldhähnchen Kommunikationsdesign
Berlin | Germany

**Cover Photo**
Michael Jaeger
Düsseldorf | Germany

**Translation**
Ken Nah, Yong Lee
Seoul | Korea

**Copy Editing**
Ken Nah, Yong Lee
Seoul | Korea

**Production**
Bernd Reinkens
gelb+ prepress & print production
Düsseldorf | Germany

**Printing**
woozoobooks
Seoul | Korea

The book is published on behalf of Design Zentrum
Nordrhein Westfalen by Red Dot Edition.

Original edition © 2017 Red Dot Edition.

Homo Ex Data by Burkhard Jacob, Vito Oražem, Peter Zec

ISBN 978-3-89939-201-2

**Red Dot Edition**
Design Publisher
edition@red-dot.de
www.red-dot-edition.com

Bibliographic information published by the Deutsche Nationalbibliothek:
The Deutsche Nationalbibliothek lists this publication in the Deutsche Nationalbibliografie; detailed bibliographic data are available on the Internet at http://dnd.ddb.de

# Homo Ex Data

The Natural of the Artificial

초판 1쇄 발행 2023년 3월 22일

글 **Peter Zec, Vito Oražem, Gretha Lösch-Schloms**
옮긴이 **나건 이용혁**
디자인 **이용혁**

펴낸이 **박현민**
펴낸곳 **우주북스**
등록 2019년 1월 25일 제2020-000093호
주소 (04735) 서울시 성동구 독서당로 228, 2층
전화 02-6085-2020
팩스 0505-115-0083
이메일 gato@woozoobooks.com
인스타그램 /woozoobooks
홈페이지 woozoobooks.com

ISBN 979-11-976863-2-0 (13500)